Plants of Significance to Bees

Stuart A. Roberts

Northern Bee Books

Plants of Significance to Bees
© Stuart A. Roberts

All rights reserved. No part of this publication may be reproduced, stored in a retrieval system, transmitted in any form or by any means electronic, mechanical, including photocopying, recording or otherwise without prior consent of the copyright holders.

ISBN 978-1-914934-47-6

Published by Northern Bee Books, 2022
Scout Bottom Farm
Mytholmroyd
Hebden Bridge
HX7 5JS (UK)

Design and artwork by DM Design and Print

Plants of Significance to Bees

Stuart A. Roberts

Alphabetical Plant Index

Page	Common Name	Latin Name	Family Name
32	Apple	*Malus pumila*	Rosaceae
137	Asparagus	*Asparagus officinalis*	Asparagaceae
66	Aubretia	*Aubretia deltaidia*	Brassicaceae
140	Barberry	*Berberis* spp	Berberidaceae
78	Bell Heather	*Erica cinerea*	Ericaceae
87	Bilberry	*Vaccinium myrtilus*	Ericaceae
25	Bird's-foot trefoil	*Lotus corniculatus*	Fabaceae
17	Black currant	*Ribes nigra*	Grossulariaceae
33	Blackberry	*Rubus frutirosus*	Rosaceae
40	Blackthorn	*Prunus spinosa*	Rosaceae
136	Bluebell	*Hyacinthoides non-scripta*	Asparagaceae
84	Blueberry	*Vaccinium corymbosum*	Ericaceae
90	Borage	*Borago officinalis*	Boraginaceae
27	Broom	*Cytisus scoparius*	Fabaceae
83	Calico bush	*Kalmia latifolia*	Ericaceae
127	Carrot	*Daucus carota*	Apiaceae
107	Catmint	*Nepeta* spp	Lamiaceae
65	Charfock	*Sinapis arvensis*	Brassicaceae
31	Cherry	*Prunus* spp	Rosaceae
34	Cherry Laurel	*Prunus lourocerasus*	Rosaceae
69	Chinese Bee Tree	*Tetradium donielli*	Rosaceae
133	Chives	*Alium schoenoprasum*	Alliaceae
95	Comfrey	*Symphytum* spp	Boraginaceae
94	Common Lungwort	*Pulmonaria officinalis*	Boraginaceae
58	Common Mallow	*Malva sylvestris*	Malvaceae
35	Cotoneaster	*Catoneaster* spp	Rosaceae
88	Cranberry	*Vaccinium oxycoccos*	Ericaceae
131	Crocus	*Crocus* spp	Iridaceae
79	Cross-leaved Heath	*Erica tetralix*	Ericaceae
114	Dandelion	*Taroxacum officinale*	Asteraceae
81	Darley Dale Heath	*Erica darleyensis*	Ericaceae
119	Dhalia	*Dhalia*	Asteraceae
21	Field bean	*Vicia faba*	Fabaceae
91	Forget-me-not	*Myosotis* spp	Boraginaceae
129	Garden Angelica	*Angelica archangelica*	Apiaceae
121	Golden rod	*Solidago* spp	Asteraceae
19	Gooseberry	*Ribes uva-crispa*	Grossulariaceae
26	Gorse	*Ulex europaeus*	Fabaceae
62	Greek mallow (Checkerbloom)	*Sidolago* spp	Malvaceae
30	Hawthorn	*Cratoegus* spp	Rosaceae
75	Himalayan balsam	*Impatiens glandulifera*	Balsaminaceae
111	Holly	*Ilex aquifolium*	Aquifoliaceae
61	Hollyhock	*Alcea rosea*	Malvaceae
53	Horse chestnut	*Aesrulus hippoacastanum*	Sapindaceae
105	Hyssop	*Hyssopus officinalis*	Lamiaceae
125	Ivy	*Hedera helix*	Araliaceae
117	Knapweed	*Centaurea nigra*	Asteraceae
101	Lavender	*Lavandula angustifolia*	Lamiaceae
109	lemon balm	*Melissa officinalis*	Lamiaceae
15	Lenten rose	*Helleborus orientalis*	Ranunculaceae
57	Lime	*Heleborus orentdis*	Malvaceae
77	Ling Heather	*Calluna vulgaris*	Ericaceae
128	Lovage	*Levisticum officinale*	Apiaceae
139	Mahonia	*Mahonia* spp	Berberidaceae
54	Maple	*Acer* spp	Sapindaceae
106	Marjoram	*Origanum majorana*	Lamiaceae
70	Mexican Orange Blossom	*Choisya ternata*	Rutaceae
100	Mint	*Mentha* spp	Lamiaceae
59	Musk Mallow	*Malva moscata*	Malvaceae
42	Musk Willow	*Salix aegypatia*	Salicaceae
64	Oilseed Rape	*Brassica napus*	Brassicaceae
37	Pear	*Pyrus* spp	Rosaceae
93	Phacelia	*Phocelia tanacetifolia*	Boraginaceae
103	Phlomis	*Phlomis fruiticosa*	Lamiaceae
86	Pieris	*Pieris* spp	Ericaceae
11	Poppy	*Papaver* spp	Papaveraceae
97	Privet	*Ligustrum vulagare*	Oleaceae
49	Purple loosestrife	*Lythrum salicaria*	Lythraceae
43	Pussy Willow	*Salix caprea*	Salicaceae
115	Ragwort	*Senecio jacobaea*	Asteraceae
36	Raspberry	*Rubus idaeus*	Rosaceae
22	Red Clover	*Trifolium pratense*	Fabaceae
18	Red Currant	*Ribes rubra*	Grossulariaceae
82	Rhododendron	*Rhododendron ponticum*	Ericaceae
51	Rosebay Willowherb	*Chamerion angustifolium*	Onagraceae
102	Rosemary	*Rosmarius officinalis*	Lamiaceae
38	Rowan	*Sorbus* spp	Rosaceae
122	Rudbeckia	*Rudbeckia fulgida*	Asteraceae
108	Sage	*Salvia officinalis*	Lamiaceae
24	Sainfoin	*Onobrychis vicifolia*	Fabaceae
118	Sea aster	*Tripolium pannonicum*	Asteraceae
73	Sea Lavender	*Limonlium vulagre*	Plumbaginaceae
104	Self-heal	*Prunella grandiflora*	Lamiaceae
134	Snowdrop	*Galanthus nivalis*	Amaryllidaceae
39	Strawberry	*Fragaria* spp	Rosaceae
85	Strawberry tree	*Arbutus unedo*	Ericaceae
116	Sunflower	*Helianthus* spp	Asteraceae
67	Sweet Alison	*Lobularia maritima*	Brassicaceae
55	Sycamore	*Acer pseudoplatanus*	Sapindaceae
120	Tansy	*Tanacetum vulgare*	Asteraceae
47	Teasel	*Dipsocus fullonum*	Caprifoliaceae
99	Thyme	*Thymus* spp	Lamiaceae
60	Tree mallow	*Lavatera*	Malvaceae
45	Verbena	*Verbena bonariensis*	Verbanaceae
92	Vipers Bugloss	*Echium vulgare*	Boraginaceae
23	White Clover	*Trifolium repens*	Fabaceae
13	Wild clematis	*Clematis vitalba*	Ranunculaceae
14	Winter aconite	*Eranthis hyemalis*	Ranunculaceae
80	Winter Heath	*Erica carnea*	Ericaceae
123	Yarrow	*Achillea millefolium*	Asteraceae

Blue shading indicates inclusion in the NDB plant list

Contents

Alphabetical plant index	Page iv
Acknowledgements, Preface & General References	Page 1
Flower Families	Page 2
NOB Plant List	Page 3
Flower Terminology	Page 4
Pollen Morphology	Page 5
Why Choose to Visit One Flower Rather than Another?	Page 8
Pollen & Nectar References	Page 10

Acknowledgements

Special thanks to Geoff Hopkinson BEM NDB for his unending support, moral guidance and friendship.

Very great thanks to Sue Townsend from the Field Studies Council (FSC) whose passion for botany helped to convince me that it was possible for me to tell my asparagus from my field beans.

And finally, many thanks to PalDat who kindly reviewed the book and gave permission for the electronmicrographs to be used. PalDat – a palynological database (2000 onwards, www.paldat.org)

Preface

This book was compiled by Stuart Roberts during 2019 and 2020. Stuart was Furloughed and this enabled him to complete the work he had started as a botany project as part of his studies.

The flower images were all taken by Stuart Roberts. The pollen photomicrographs were all taken from the pollen samples collected during the compilation of this book by Stuart Roberts. The images were taken with a digital camera attached to an Olympus BH2 biological microscope at x400 magnification.

General References

1. The electronmicrographs of the pollen grains are all from https://www.paldat.org/
2. The honey source references are from Eva Crane's 'Directory of Important World Honey Sources' 1984. Published by Northern Bee Books. Note : the reference in the text to 'Nectar Presentation' is based on honey sources in the UK only.
3. Plant family information was from 'Botany in a Day' by Thomas J. Elpel 6th edition 2018. Published by Hops Press
4. Plant information was from 'Plants for Bees' by Kirk and Howes 2012. Published by IBRA.
5. Pollen morphology was from :
    a) "An Illustrated Guide to Pollen Analysis' by Moore & Webb 1978. Published by Hodder and Stoughton Educational.
    b) 'Pollen Morphology and Plant Taxonomy - Angiosperms' by G. Erdtman 1952. Published by Almquist & Wiksells.
    c) 'Pollen Microscopy' by Norman Chapman 2nd edition 2018. Published by CMI Publishing.

Please note there are some specific pollen and nectar references on page 11.

Flower Families

Family Name   Page Number

Papaveraceae 10
Ranunculaceae 12
Grossulariaceae 16
Fabaceae 20
Rosaceae 28
Salicaceae 41
Verbenaceae 44
Caprifoliaceae 46
Lythraceae 48
Onagraceae 50
Sapindaceae 52
Malvaceae 56
Brassicaceae 63
Rutaceae 68
Plumbaginaceae 72
Balsaminaceae 74
Ericaceae 76
Boraginaceae 89
Oleaceae 96
Lamiaceae 98
Aquifoliaceae 110
Asteraceae 112
Araliaceae 124
Apiaceae 126
Iridaceae 130
Alliaceae 132
Amaryllidaceae 134
Asparagaceae 135
Berberidaceae 138

NDB Plant List

The NDB board publish a list of plants "the NDB Plant List". It was last updated (at the time of writing) in January 2015.

It states :

Plants are typically selected from this list for the NDB Practical Examination. However, this list is provided as a guide only; any plant useful or significant to bees could be included. It should be noted that the botany section of the NDB exam is not concerned exclusively with plant identification; aspects of pollen morphology, nectar presentation, preferred habitat and particular special features may also feature.

Family	Plants
Papaveraceae	Poppy (*Papaver spp*)
Ranunculaceae	Wild clematis (*Clematis vitalba*)
Grossulariaceae	Black and Red currants (*Ribes nigra* and *R. rubra*), Gooseberry (*R. uva-crispa*)
Fabaceae	Field bean (*Vicia faba*), Red & White clover (*Trifolium pratense & T. repens*), Sainfoin (*Onobrychis viciifolia*), Bird's-foot trefoil (*Lotus corniculatus*)
Rosaceae	Hawthorn (*Crataegus spp*), Cherry (*Prunus spp*), Apple (*Malus pumila*), Blackberry (*Rubus fruticosus*), Cherry laurel (*Prunus laurocerasus*), Cotoneaster spp, Raspberry (*Rubus idaeus*)
Salicaceae	Willows (*Salix spp*)
Lythraceae	Purple loosestrife (*Lythrum salicaria*)
Onagraceae	Rosebay willowherb (*Chamerion angustifolium*)
Sapindaceae	Horse chestnut (*Aesculus hippocastanum*), Maple (*Acer spp*), Sycamore (*A. pseudoplatanus*)
Malvaceae	Lime (*Tilia spp*)
Brassicaceae	Oilseed rape (*Brassica napus*)
Plumbaginaceae	Sea lavender (*Limonium vulgare*)
Balsaminaceae	Himalayan balsam (*Impatiens glandulifera*)
Ericaceae	Heather & Heaths (*Calluna & Erica spp.*) Rhododendron (*Rhododendron ponticum*)
Boraginaceae	Borage (*Borago officinalis*), Forget-me-not (*Myosotis spp*), Vipers Bugloss (*Echium vulgare*), Phacelia (*Phacelia tanacetifolia*)
Oleaceae	Privet (*Ligustrum vulgare*)
Lamiaceae	Thyme (*Thymus spp*), Mint (*Mentha spp*), Lavender (*Lavandula angustifolia*), Rosemary (*Rosmarinus officinalis*)
Aquifoliaceae	Holly (*Ilex aquifolium*)
Asteraceae	Dandelion (*Taraxacum officinale*), Ragwort (*Senecio jacobaea*), Sunflower (*Helianthus spp*); Knapweed (*Centaurea nigra*), Sea aster (*Aster tripolium*)
Araliaceae	Ivy (*Hedera helix*)
Apiaceae	Carrot (*Daucus carota*)
Iridaceae	Crocus spp
Alliaceae	Chives (*Allium schoenoprasum*), Snowdrop (*Galanthus nivalis*)
Asparagaceae	Bluebell (*Hyacinthoides non-scripta*), Asparagus (*Asparagus officinalis*)

This book has been put together as a set of study notes to address the following :

1. Plants of significance to bees based on the the NDB plant list
2. The pollen morphology of those plants
3. The nectar presentation of those plants
4. The preferred habitat of those plants
5. Any special features of those plants

I have tried to make comments on all of the features listed above for each of the plants chosen to be included in the book.

Note - the list includes both UK native and non-native plants.

Flower & Plant Terminology

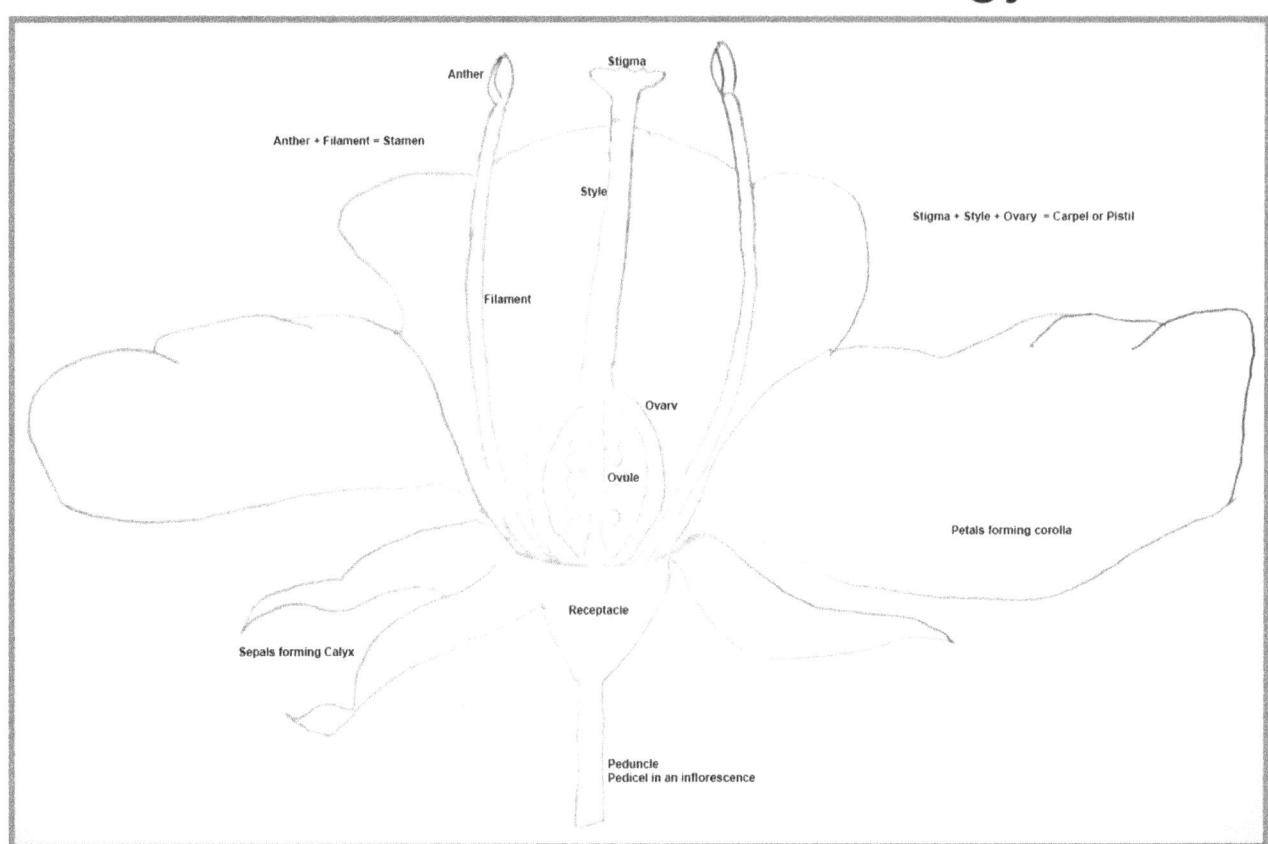

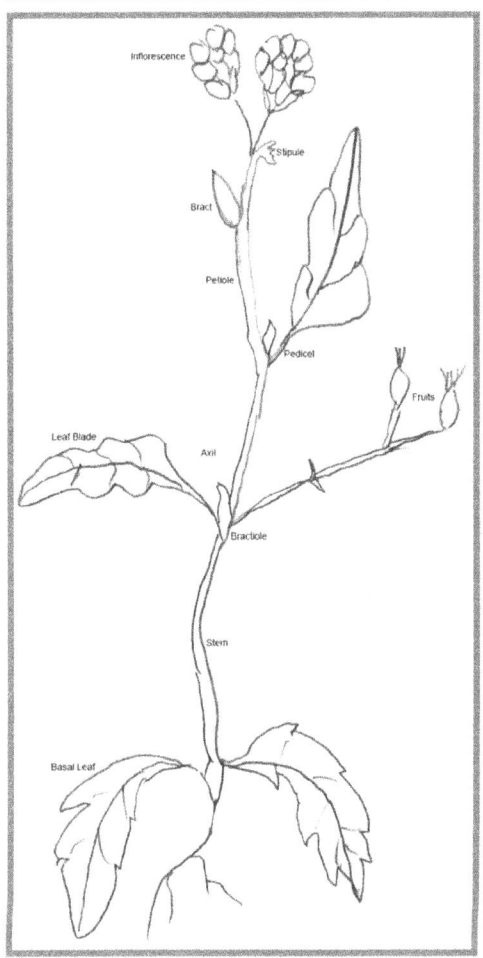

Flower Part Definitions

- Anther - the distal end of the stamen where pollen is produced
- Filament - the stalk of a stamen
- Stamen - the male organs
- Stigma - the expanded tip of a column arising from the ovary
- Style - a column arising from the ovary
- Ovary - the structure that contains the ovules
- Carpel - the female organs
- Pistil - the ovary, stigma and style
- Ovule - structure that develops into a seed when fertilised
- Petals - modified leaves that surround the reproductive parts of flowers
- Corolla - the collection of petals of a flower
- Receptacle - the end of the pedicel that joins to the flower
- Peduncle - the part of a stem that bears the entire inflorescence
- Sepals - non-reproductive structures involved in protecting the flower when it is still a bud
- Calyx - the outer whorl of sepals

Plant Part Definitions

- Inflorescence - a cluster of flowers
- Stipule - paired scales, spines, glands, or blade-like structures at the base of a petiole
- Bract - a modified leaf usually beneath a flower
- Petiole - a leaf stalk supporting a blade and attaching to a stem at a node
- Pedicel - the stem or stalk that holds a single flower in an inflorescence
- Leaf blade - the main body of the leaf
- Axil - the angle between the leaf and the stem
- Fruit - a new entity containing dormant seeds
- Bracteole - a secondary smaller bract
- Stem - the plant axis that bears buds and shoots with leaves and, at its basal end, roots
- Basal leaf - a leaf at the base of the plant

Pollen Morphology

Pollen

The male gametes of plants are contained within pollen grains. The pollen grain has an extremely durable outer coating that enables pollen to withstand the harshest environmental conditions. The end goal is to get the pollen grain from the anther of the pollen producing plant to the stigma of the receiving plant. When the pollen grain lands on the stigma of the receiving plant a chemical reaction occurs and this initiates the growth of the pollen tube from within the pollen grain. The pollen tube will grow down the style of the receiving plant until it reaches the ovary and fuses with it. At the ovary, the pollen tube facilitates the passage of the sperm cell from the pollen grain to the ovule of the plant which contains the egg cell. The coming together of these two cells is called fertilisation.

The importance of pollen to bees

The collection of pollen by honey bees is to supply their main source of protein. There is a small amount of protein in honey but pollen has far more. Not all pollens are equal. There is a variation in the amount of protein in pollens and there is also a difference in the amino acid content too. Honey bees need 10 essential amino acids. These are 'essential' because the honey bee cannot synthesise these form other amino acids. It is, therefore, important for the honey bee to have access to plants that supply these amino acids.

The importance of pollen to the beekeeper

The pollen collected by the bees has fascinated people for a long time. It is obvious to anyone who takes time to look, by the colours of the pollen loads, that there are a variety of pollens collected. Identification of the pollen collected by the bees enables the beekeeper to understand what forage plants the bees are visiting. Identification of pollen in honey enables the beekeeper to identify the nectar sources too. It may then be possible to label a product as a 'single-source' honey such as Borage Honey or Heather Honey. There is a complication in that certain pollens naturally present themselves more than others and so may be over represented in honey. This must be taken into account before any claim of single-source honey can be made.

Features that enable pollen identification

To be able to successfully identify individual pollens, which is not as easy as it sounds, we need to be able to describe what we see down the microscope. There are certain features that enable the description of pollen grains and they are as follows :

The easy ones are :

- Colour - but this assumes that the pollen was seen in large quantities either on the flower, the bee or packed in a cell.

- Size - it is relatively easy to measure the size of pollen grains, under a microscope, using a graticule or you can compare to a pollen of known size such a Hazel which is 25 microns in diameter (0.000025 m)

- Shape - pollen grains come in a variety of different shapes and sometimes pollen grains are grouped together - see Ling Heather - slide 49 page 77.

More difficult features are :

- Apertures - these are holes in the exine that are where the pollen tube emerges during germination when the pollen grain has reached a compatible stigma. There are 2 types :

    o   Colpi (furrows) - more primitive than pori and are long and boat shaped with pointed ends
    o   Pori (pores) - are essentially round holes but may be slightly elongated with rounded ends

- Exine structure - this is sometimes called sculpturing.

To describe this in any detail requires some knowledge of the structure of the pollen. This following diagram shows a section through a pollen grain to display the various layers.

A few definitions of terms to describe the exine structure are :

Tectum - 'Roof' layer which joins the tops of the columellae - forming the outer layer of the sexine

Columella - Term for small rod-like elements that protrude radially and for the inner layer of the sexine

Nexine (1 and 2) - Inner unsculptured part of the exine which appears solid or sometimes layered

Sexine - Outer sculpted part of the exine (tectum + columellae = sexine)

These terms become important in being able to describe the pollen structure under the microscope and we will see later how the various elements lead to different patterns on the pollen surface.

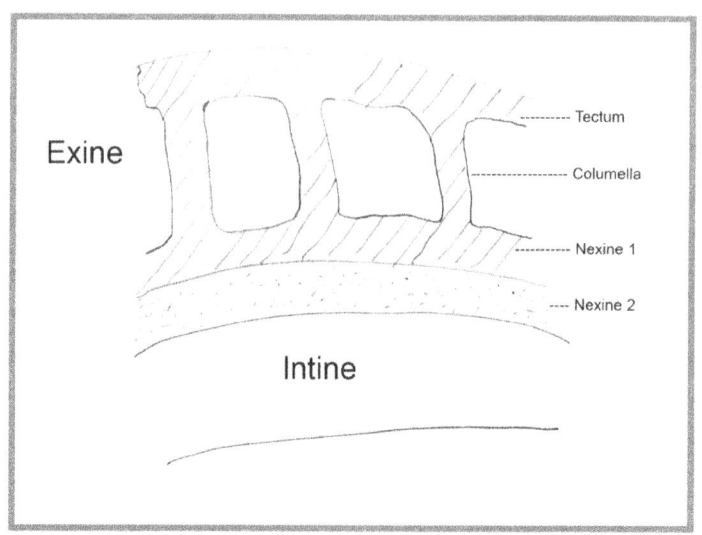

This, however, still presents a problem with how to describe where the features are on the pollen grain. So the accepted convention is that the terms 'polar' and 'equatorial' are used to enable more specific descriptions. For example :

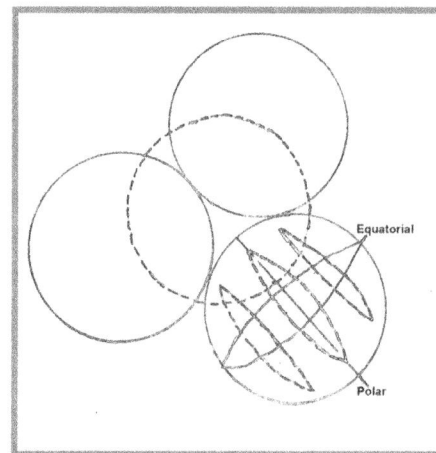

To the left is a diagram of a tetrad (four pollen grains in a tetrahedral formation). The diagram is labelled to show how the poles and the equator can be used to describe pollen features.

It also shows colpi or furrows that cross the equatorial plane extending towards the poles of the pollen grain.

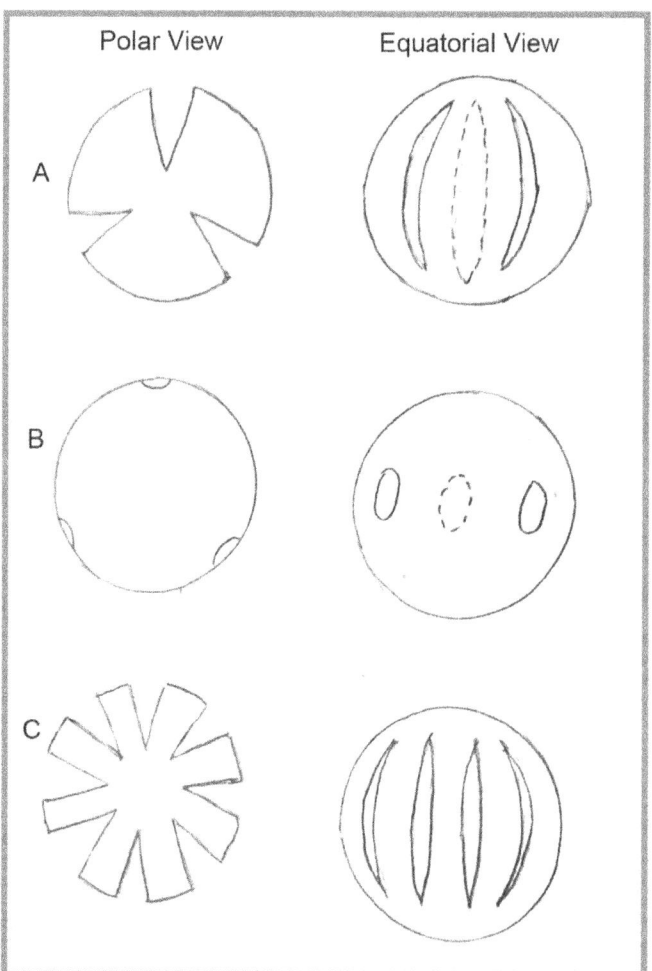

Using the descriptors gives the following :

Consider pollen A : the shape is spherical and is has 3 colpi or furrows and so the pollen can be describeed as colpate (with furrows).

An example of a plant with pollen like this is Acer spp - the maples.

Pollen B - is again spherical but this time has 3 pores. The pollen can be described as porate (with pores).

The example here is Betula spp - birch.

Pollen C - is again spherical and colpate but this time has 8 furrows around the pollen grain.

The example here is Nepeta spp - catmint.

This is not the whole story. Next we must consider the surface patterning which is a direct result of the structure of the exine.

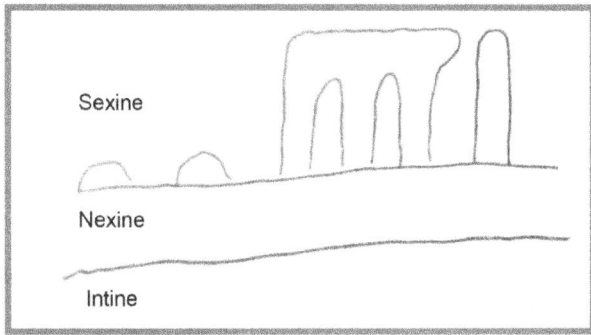

The surface of pollen grains is textured as a result of the structure of the exine, specifically the sexine. I previously drew a section through the exine but showing a very simplified structure. Actually there are numerous combinations or possibilities that result in varying patterns seen on the surface of the pollen grains. A few examples are as follows :

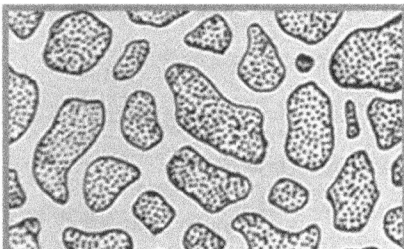

To the left is the pattern seen on the surface of the pollen. To the right are the structures that result in this pattern. The structures are described as tectate and semi-tectate.

Tectate means with a complete 'roof' and semi-tectate is with a partial 'roof'.

The pattern itself is called 'reticulate' which means projecting elements arranged in a network pattern.

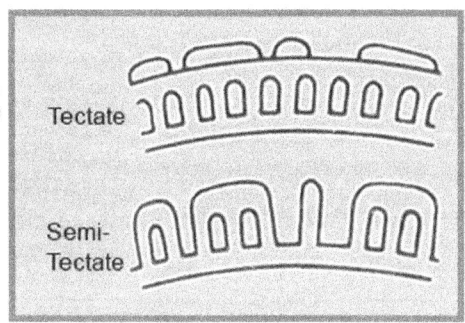

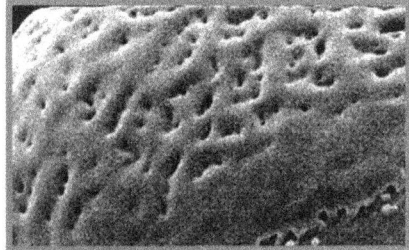

This image is an electron-micrograph of a trifolium - clover pollen surface showing the reticulate pattern from a tectate sexine.

This image is an electron-micrograph of a salix or willow pollen surface showing the reticulate pattern from a semi-tectate sexine.

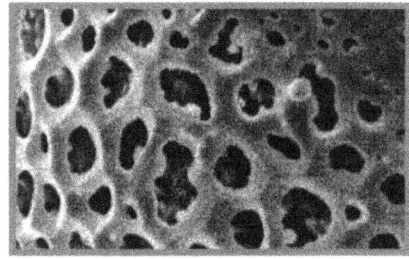

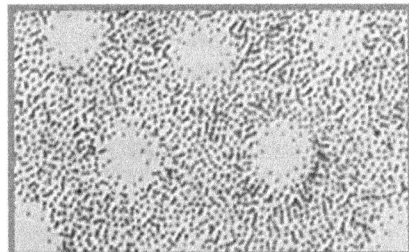
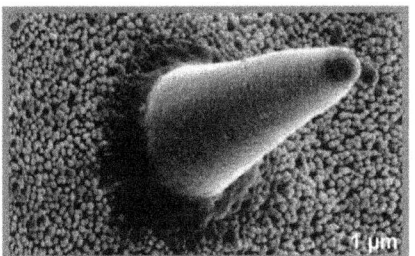

Here the pattern is called echinate and the example pollen is a malva or mallow.

It is a tectate structure as is shown here.

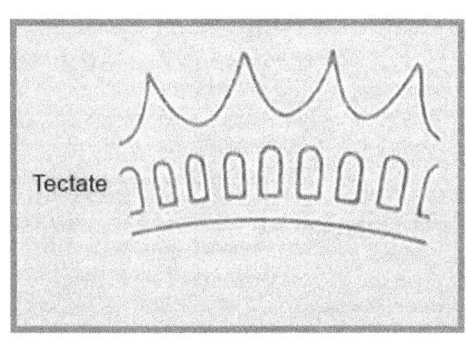

Why Choose to Visit One Flower Rather than Another ?

There is an assumption that bees choose which flowers to visit based on knowledge of the potential reward to be had. The honey bee obtains nectar and pollen from these sources so the assumption is that there is a relationship between the constituents of nectar, pollen or both and the attractiveness of the flowers. Once a flower species has been chosen then successive visits by individual bees and nest mates will be made. However, bees leave pheromone markers on flowers that they have visited to prevent other bees visiting flowers that have had the nectar drained.[1]

I am more interested in why a bee would choose to forage on dandelions rather than oil seed rape or vice versa. There has been work done on flower shape to determine if it is simply the shape of the flowers that determine where a bee forages.[2]

Work has been ongoing since the early 20th century on the nectar and pollen constituents. In particular the work of G. R. Wykes from the Bee Research Department, Rothamsted Experimental Station. Wykes built upon the work of Karl von Frisch and Colin Butler but couldn't determine a connection between nectar constituents and the foraging habits of bees. Seeley took this work and expanded it to include a series of equations that balanced the calorific income of the nectar against the energy expanded in bringing the nectar back to the nest.[10] Seeley terms this the 'currency of patch quality' and concludes that the bees 'calculate' net energy efficiency.

Nectar

- Water content - the water content of nectar is dependent upon environmental factors. The water in nectar is evaporated in hot environments and nectar is diluted by rain or atmospheric moisture in humid environments. This also depends upon the flower morphology. If a flower is bell-shaped and hanging down then it is less likely to be affected by rain than an open upward-facing flower.

- Sugar content - the main solutes in nectar are glucose, fructose and sucrose and their content ranges from between 7 and 70 %. Much work has been done to determine if the ratios of the sugars affect the attractiveness of the flowers to pollinators. This was led by G. R. Wykes , continued by Herbert and Irene Baker in 1983 and furthered by Kevan later. The sugars come from the phloem sap, the products of synthesised nectary starch or degradation of nectary parts.

- Amino acids and proteins - This is the biggest group of substances in nectar after the sugars. The amino acids in nectar can be both essential and non-essential and the proteins are enzymes and preservatives. These amino acids and proteins are derived from the phloem sap, the products of synthesised nectary starch or degradation of nectary parts. Nectar amino acids may play a part in the taste preferences of insects.[4]

- Nectaries[11] - are either extra-floral or floral
  - Floral nectaries:
    - Reward animals transporting pollen
    - Are in separate parts of the plant
    - Are visited by insects, birds and mammals
    - Secrete for a few hours to several days
    - Produce up to a few millilitres per day
    - Have variable nectar quality
  - Extra-floral nectaries:
    - Rewards animals defending the plant from herbivores
    - Can be in the leaf, the petiole, the stipule or the blade
    - Are visited by ants mainly
    - Secrete for a few days to a few months
    - Produce only microlitres per day
    - Have very common, unchanging nectar for ants mainly

- Nectar Standing Crop - this is defined by Kearns and Inouye (1993)[3] as "the quantity and distribution of nectar determined by randomly sampling flowers that have not been protected from pollinators". There is symbiotic relationship between nectar standing crop and animal visits. The standing crop is affected by animal visits and animal visits are affected by the standing crop. It could be that a plant that has a large standing crop is more attractive than one that has a small standing crop.

- Nectar production - it is accepted that the amount of nectar produced by plants is linked to the requirements of the pollinating animal though co-evolution. A plant that is pollinated by a high energy pollinator like a bat generates a lot of nectar compared to a plant that is pollinated by bees or butterflies.[5] Could bees know instinctively which flowers to visit because of co-evolution?

- Nectar production rate - Cruden[5] recognised 3 classes of nectar production rate those being slow, fast and super producers. There is also a correlation between production rate and the protection of the nectaries. Most nectar production take place before foraging visits have begun. Most plants, therefore, have storage capacity for nectar and this tends to be protected in slow producers but less so in fast or super producers. Nectar cessation can occur for 2 reasons. Firstly, the nectar storage has reached a maximum. Secondly that foraging has stopped. In the first instance, if the nectar is removed then production restarts. We accept that bees visit slow, fast and super producers and foraging activity is governed by the weather amongst other things.

Pollen

It is known that pollen is the main source of protein in the diet of the honey bee. It is also known that as a bee ages there is a different dietary requirement and at foraging age, circa, 3 weeks, the worker bee diet is almost exclusively nectar.[6] There is direct relationship between pollen availability and brood rearing in honey bees. In the spring when there is an abundance of pollens the brood rearing is at a maximum and in the summer when pollen availability reduces then the brood production rate reduces too.[7] Attractiveness of plants in terms of their pollen constituents has been examined but very little data is available. However, there is a possible link to the lipid content of the pollen with the attractiveness of the flower to honey bee foragers.[8] The Brassicaceae being particularly high in pollen lipids and particularly attractive to honey bees. However, this might not be the full story. D.C. Somerville found that a reliable measure of pollen quality was crude pollen content.[9] He loosely categorised various pollens into poor, average, above average and excellent categories - see the table below. This data formed the basis for the paper "Fat Bees, Skinny Bees".

	Crude Protein
Minimum Ideal %	20.0
Viper's Bugloss	34.9
Sunflower	12.9 - 18.5
Lavender	19.4
Pear	26.2
Blackberry	14.8 - 20
Pussy Willow	21.9
Ragwort	11.8 - 17.3
White Clover	22.5 - 25.9
Gorse	16.5 - 28.4
Blueberry	13.9
Field Beans	22.3 - 24.4
Almond	23.3 - 30.7
Asteraceae	14.5 - 24.5
Oilseed Rape	22.8 - 27.1
Raspberry	21.3

Pollen Class

Viper's Bugloss is classed as excellent (above 30%)
The dark green are classed as above average (25-30%)
The light green are classed as average (20-25%)
The red are classed a poor (below 20%)

Lipid Content
Ideal minimum is not identified
The higher the better = more attractive

	Fat / Lipid
Minimum Ideal %	Not known
Viper's Bugloss	4.1
Sunflower	1.1 - 1.4
Lavender	2.9
Pear	1.8
Pussy Willow	3.1
Ragwort	2.4
White Clover	2.5
Gorse	2.1
Blueberry	2
Field Beans	1.72
Almond	1.89 - 2.74
Asteraceae	Circa 7.2
Oilseed Rape	6.8 - 7.3

- Lipid content - D. C. Somerville wrote a paper for NSW Agriculture (Australia). Unfortunately, a lot of the plants surveyed do not appear in this book. However, there was a general trend in that the higher the pollen lipid content the more attractive the plants appeared to be to bees. See the table above (right hand side) for the lipid contents of the plants within this book.

Amino Acid	Threonine	Valine	Methionine	Leucine	Isoleucine	Phenylalanine	Lysine	Histidine	Arginine	Tryptophan
Minimum Ideal Ratio	3.0	4.0	4.5	4.5	4.0	2.5	3.0	1.5	3.0	1.0
Viper's Bugloss	4.7	5.5	2.3	7	4.6	4.3	5.1	2.4	4.9	-
Sunflower	4.2	5	2.3	6.8	4.6	4.6	6.2	4.8	4	-
Lavender	4.2	4.5	2.2	6	3.6	4.1	6.4	3.7	4.3	-
Pear	4.2	4.5	2.2	6	3.6	4.1	6.4	3.7	4.3	-
Blackberry	4.4	5.4	2.3	7.3	4.6	4.6	8.5	2.6	7.4	0.9
Pussy Willow	4.5	5.5	2.5	7.5	4.8	4.4	7.2	2.3	6.3	-
Ragwort	4.3	4.2	2.1	6.2	3.5	4	6.2	4.1	6	-
White Clover	4.6	5.3	2.2	7	4.4	4.3	5.9	2.5	4.7	-
Gorse	4.5	5.1	2.3	7.2	4.4	4.4	6	2.3	4.7	-
Blueberry	3.8	5.4	2.3	6.7	4.7	3.5	6.4	2	5.6	-
Field Beans	4.6	5.2	2.2	6.7	4.8	4.2	6.2	2.1	5.1	-
Almond	4.5	5.4	2	6.7	4.1	4.9	6	2.1	5.3	1.1
Oilseed Rape	4.9	5.1	2.3	7	4.6	4.3	8.2	2.1	5.1	-

Red shading indicated a deficiency in the appropriate amino acid

- Amino acid content - there are about 500 naturally occurring amino acids and we know that bees need 10 essential amino acids (EAAs) and that not many pollens offer all 10. This would potentially drive a colony need to forage for a wide range of pollens to try to achieve a reasonable quantity of each of the 10 EAAs.

Papaveraceae - Poppy Family

Key words : Petals in fours with numerous stamens and often milky sap

The Papaveraceae are an economically important family of about 42 genera and approximately 775 known species of flowering plants.

The family is cosmopolitan, occurring in temperate and subtropical climates (mostly in the northern hemisphere), but almost unknown in the tropics.

Most are herbaceous plants, but a few are shrubs and small trees. Common traits are milky sap and altenate leaves. The flowers are regular or bisexual with 2or 3 sepals which fall away as the flowers open. They have a superior ovary and consists of at least 2 fused carpels indicated by the number of stigmas above. The carpels fuse to form a single chamber which eventually becomes a capsule containing many seeds.

Species of importance to bees :

1. Common Name : Poppy
   Latin Name : **Papaver** spp

Poppy field near the National Memeorial Arboretum, Alrewas, Staffordshire

Pollen & Nectar References

1. Can Bees Select Nectar-Rich Flowers in a Patch? - Sarah A. Corbet,C. J. C. Kerslake,D. Brown &N. E. Morland - Pages 234-242 Received 05 Mar 1984, Published online: 24 Mar 2015.
2. Honeybees prefer novel insect-pollinated flower shapes over bird-pollinated flower shapes - Scarlett R Howard, Mani Shrestha, Juergen Schramme, Jair E Garcia, Aurore Avarguès-Weber, Andrew D Greentree, Adrian G Dyer - Current Zoology, Volume 65, Issue 4, August 2019, Pages 457-4653.
3. Kearns, C.A. & Inouye, D.W. (1993). Techniques for Pollination Biologists. pp. 153-215. Colorado: University Press of Colorado.
4. The taste of nectar - a neglected area of pollination ecology Mark C. Gardener and Michael P. Gillman, Dept of Biological Sciences, The Open Uni.,Forum.
5. Cruden, R. W., Hermann-Parker, S. M., and Peterson, S. 1983. Patterns of nectar production and plant pollinator coevolution, in "Biology of Nectaries" (T. S. Elias and B. A. Bentley, Eds.), pp. 81-125, Columbia Univ. Press, New York.
6. Nutritional balance of essential amino acids and carbohydrates of the adult worker honeybee depends on age - Pier P. Paoli, Dion Donley, Daniel Stabler, Anumodh Saseendranath, Susan W. Nicolson, Stephen J. Simpson, Geraldine A. Wright.
7. The Impact of Different Protein Content of Pollen on Honey Bee (Apis mellifera L.) Development - Zheko Radev - American Journal of Entomology 2018; 2(3): 23-27.
8. Lipid content of honey bee-collected pollen from south-east Australia, D. C. Somerville, Australian Journal of Experimental Agriculture, 2005, 45, 1659-1661.
9. Nutritional Value of Bee Collected Pollens - A report for the Rural Industries Research and Development Corporation by DC Somerville, May 2001, RIRDC Publication No. 01/047, RIRDC Project No. DAN-134A
10. Social Foraging by Honeybees: How Colonies Allocate Foragers among Patches of Flowers - Thomas D. Seeley - Behavioral Ecology and Sociobiology, Vol. 19, No. 5 (1986), pp. 343-354.
11. Nectaries and Nectar - Susan W. Nicolson, Massimo Nepi, Ettore Pacini, Springer Science & Business Media, 18 Apr 2007.

Poppy

Family : Papaveraceae (Poppy Family)
Common Name : Poppy

Latin Name : *Papaver* spp
Alternate Names : None

Description : Annuals with 1-2 pinnate leaves

Flowering times : June - September

Flowers : flowers of species have 4 to 6 petals, many stamens forming a conspicuous whorl in the centre of the flower. The petals are showy, may be of almost any colour and some have markings.

Leaves : poppies have lobed or dissected leaves

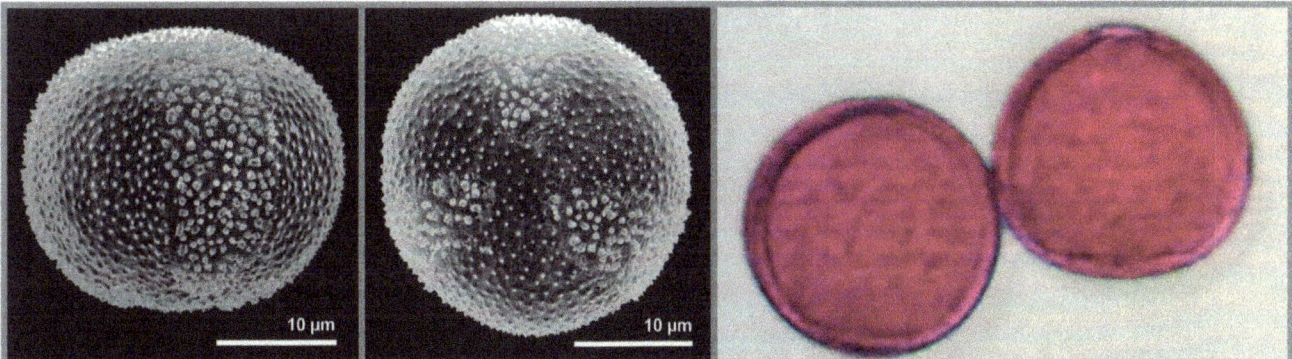

Pollen : Pollen loads are dark blue to black. The pollen grains are 25 um in diameter. There are 3 apertures, in this case, furrows and the exine appears quite thin. The surface structure is granular as is the surface structure of the furrows.

The poppy produces pollen in abundance which attracts numerous species of bees including honey bees.

Nectar presentation : No honey crop. Which is a shame because the honey might well have had a narcotising effect. Although, some people have reported that there might be nectar production at times - one report suggested that honey bees returning from poppies (pollen loads vsible) had trouble finding the entrance to the hive.

Preferred habitat : Arable fields, disturbed ground, road-side verges especially on calcareous soils. They like open soil to germinate and do not like competition. The seeds can be dormant in the soil for a long time. They like well drained soil and are drought tollerant.

Particular special features : Poppies include many ornamental flowers plus P. somniferum from which we obtain opium, morphine, heroin and codeine. Poppies also have edible seeds used for seasoning. The latex like sap is also a narcotic but potency varies from species to species.

SEM pictures by Oberschneider, W. In: PalDat - A palynological database.
https://www.paldat.org/pub/Papaver_rhoeas/303904; accessed 2022-03-31 Polar and Equatorial

Ranunculaceae - Buttercup Family

Key words : Dicot flowers with 3 or more simple pistils, usually with hooked tips

Ranunculaceae is a family of over 2,000 known species of flowering plants in 43 genera, distributed worldwide.

The largest genera are Ranunculus (600 species), Delphinium (365), Thalictrum (330), Clematis (325), and Aconitum (300).

Ranunculaceae are mostly herbaceous annuals or perennials, but some woody climbers or shrubs.

Look for multiple simple pistils usually with hooked tips.

Species of importance to bees :

1. Common Name : Clematis
   Latin Name :   *Clematis vitalba*

2. Common Name : Winter aconite
   Latin Name :   *Eranthis hyemalis*

3. Common Name : Lenten rose
   Latin Name :   *Helleborus orientalis*

Winter Aconite in woodlands

Clematis

Family : Ranunculaceae (Buttercup Family)
Common Name : Clematis

Latin Name : *Clematis vitalba*
Alternate Names : Traveller's Joy or Old Man's Beard

Description : Annuals with 1-2 pinnate leaves

Flowering times : July - August

Flowers : white in colour, the various stamens are clearly visible and the flowers are in clusters. The flowers are around 2 cm in width.

Leaves : pinnately compound, with three to five leaflets which are elliptical in shape with rough toothed margins.

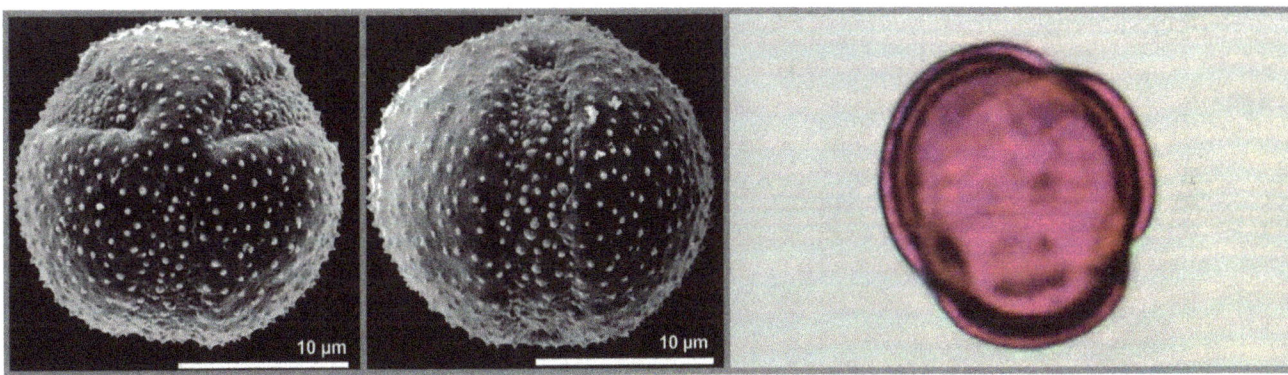

Pollen : Pollen loads are yellow. The pollen grains are 20 um in diameter. There are 3 apertures, in this case, furrows and the exine appears of medium thickness. The surface structure is granular as is the surface structure of the furrows.

Nectar presentation : No honey crop. However is does produce nectar in a peculiar way. The nectar is offered on the fillaments and not from the nectaries.

Preferred habitat : Scrub, hedgerows, wood margins and railway embankments. It prefers calcareous soils. They prefer to be in sun or partial shade and like a well drained soil.

Particular special features : It is native and widespread in the UK. It is visited by bumblebees (short and long tongued) solitary bees and honey bees.

Note - there are lots of garden varieties - many of which flower in winter.

SEM pictures by Oberschneider, W. In: PalDat - A palynological database.
https://www.paldat.org/pub/Clematis_vitalba/303872; accessed 2022-03-31 Polar and Equatorial

Winter aconite

Family :	Ranunculaceae (Buttercup Family)	Latin Name :	***Eranthis hyemalis***
Common Name :	Winter aconite	Alternate Names :	Winter hellebore
Description :	Small tuberous perennials	Flowering times :	January - March

Flowers : large 2-3 cm, yellow, cup-shaped flowers held above a collar of 3 leaf-like bracts

Leaves : palmately or pinnately lobed basal leaves and cup-shaped flowers held above a collar of deeply lobed stem leaves.

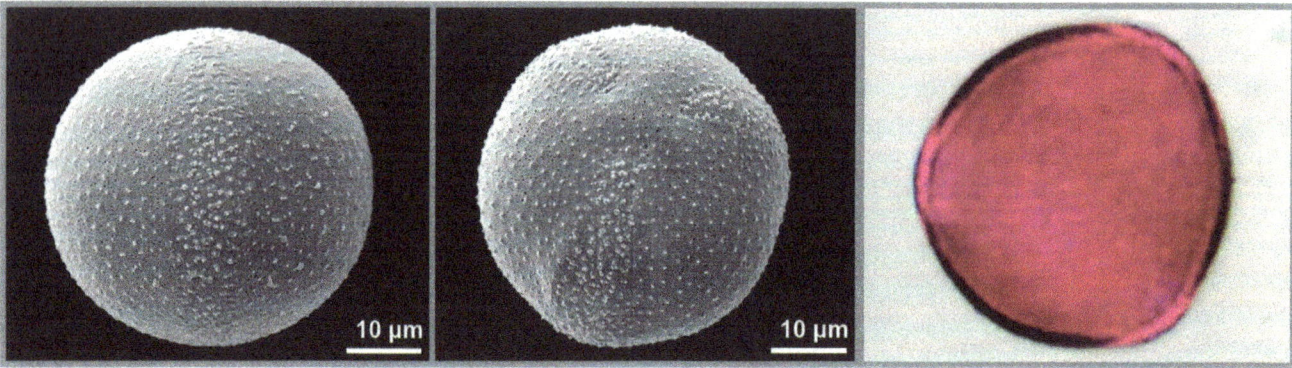

Pollen : Pollen loads are yellow. The pollen grains are 40 um in diameter. There are 3 apertures, in this case, furrows and the exine appears quite thin. The surface structure is granular as is the surface structure of the furrows.

Nectar presentation : No honey crop.

Preferred habitat : Parks and woodland. They like partial shade and a humus rich soil, moist but well drained.

Particular special features : A good early pollen and nectar provider for bees. They only open their flowers once the temperature gets above 10 degrees Celsius. This is a good thing because the bees will be foraging at 10 degrees. The nectar is contained in nectaries that are modified petals.

Not to be confused with Aconite (Monk's hood) which is a good plant for bumblebees but is not a good honey bee plant.

SEM pictures by Halbritter, H. In: PalDat - A palynological database.
https://www.paldat.org/pub/Eranthis_hyemalis/303852; accessed 2022-03-31 Polar and Equatorial

Lenten rose

Family : Ranunculaceae (Buttercup Family) Latin Name : **Helleborus orientalis**
Common Name : Lenten rose Alternate Names : Christmas rose

Description : Perennial elegant garden plants Flowering times : February - April

Flowers : pretty, pendent or outward-facing, saucer-shaped flowers.

Leaves : they have no true leaves on their flower stalks (although there are leafy bracts where the flower stalks branch).

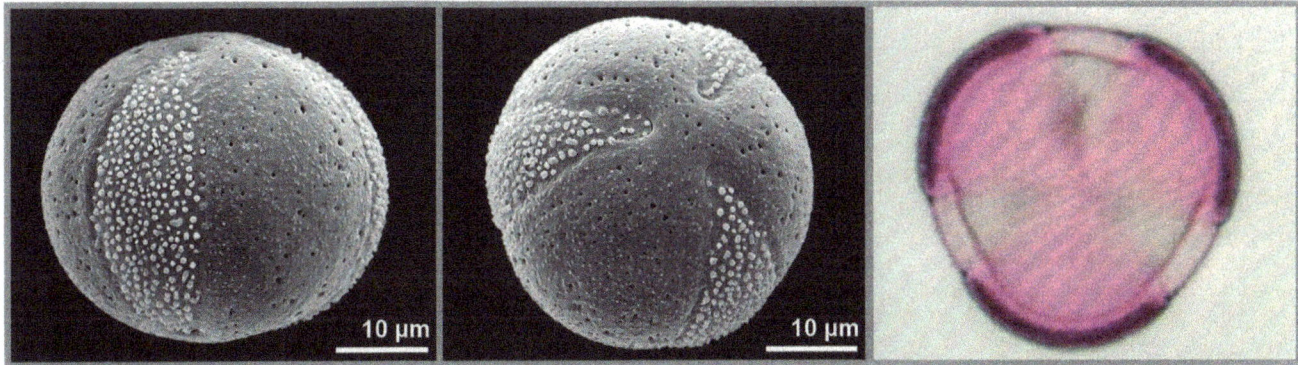

Pollen : Pollen loads are yellow. The pollen grains are 30 um in diameter. There are 3 apertures, in this case, furrows and the exine appears of medium thickness. The surface structure is net or pitted whereas the surface structure of the furrows is granular.

Nectar presentation : No honey crop.

Preferred habitat : They are found in woodlands on damp, calcareous soils.

Particular special features : Again the nectar is contained within modified petals for nectaries. What appear to be the petals are actually modified, coloured sepals.

Hellebores contain cardiac glycosides which stimulate heart contractions and should not be consumed as they are potentially very dangerous.

Due to recent taxonomic changes Helleborus orientalis is also called Helleborus hybridus. The most common native is Helleborus foetidus which flowers in early spring and attracts lots of bees.

SEM pictures by Halbritter, H. In: PalDat - A palynological database.
https://www.paldat.org/pub/Helleborus_niger/306248; accessed 2022-03-31 Polar and Equatorial

Grossulariaceae - Gooseberry & Currants

Key words : Flowers with multiple stamens fused in a central column

Worldwide, there is only 1 genus consisting of 150 species

They have regular, bisexual flowers, usually about 8 mm in diameter. The blossoms are yellow, white, greenish or sometimes red. The flowers have 5 united sepals and 5 separate petals.

There are 5 stamens alternate with the petals.

The ovary is inferior with 2 carpels as indicated by the 2 styles. The fused carpels matures as a berry with several to numerous seeds.

They have very distinctive leaves

Species of importance to bees :

1. Common Name : Black Currants
   Latin Name : ***Ribes nigra***

2. Common Name : Red Currants
   Latin Name : ***Ribes rubra***

3. Common Name : Gooseberry
   Latin Name : ***Ribes uva-crispa***

Andrena fulva - Tawny Mining Bee foraging on red currants

Black Currant

Family : Grossulariaceae (Currant Family)
Common Name : Black Currants

Latin Name : *Ribes nigra*
Alternate Names : None

Description : A medium-sized shrub

Flowering times : April - May

Flowers : inconspicuous, regular, approximately 8 mm across, reddish- or brownish green. Calyx five-lobed, wheel-shaped, hairy, with yellow glands.

Leaves : alternate with fairly long stalks, strong-scented. Blade with palmate venation, 3-5-lobed, with cordate base, toothed margins.

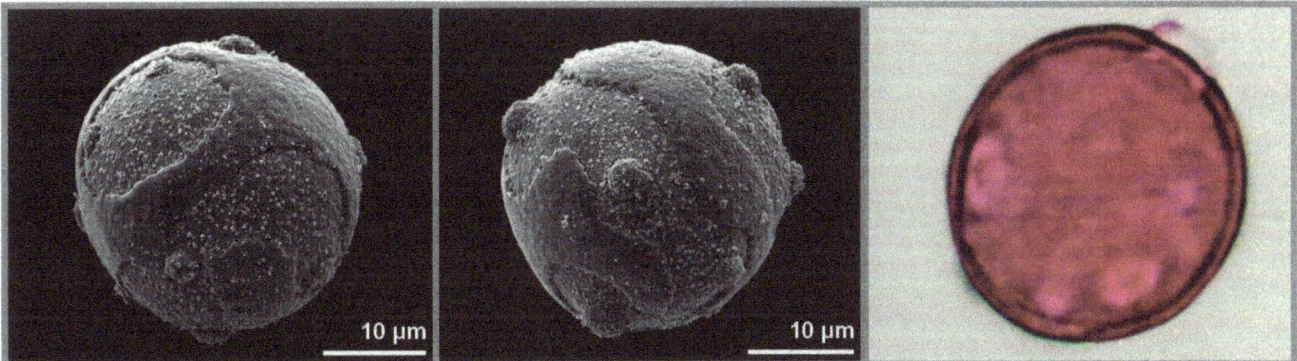

Pollen : Pollen loads are yellow. The pollen grains are 30 um in diameter. There are multiple apertures, in this case, pores and the exine appears of medium thickness. The surface structure of the exine is smooth.

Nectar presentation : No honey crop.

Preferred habitat : Damp woodland fen. They enjoy sun or partial shade but do best in fertile, well drained, moisture retentive soils.

Particular special features : Often visited by queen bumblebees, they use 'buzz' polination to extract pollen from the anthers. These plants are good for both nectar and pollen. Black currants are visited by short and long-tongued bumblebees, honey bees and solitary bees.

When the leaves are crushed the smell very strongly of black currants.
The seeds contain omega-3 and omega-6 fatty acids.

SEM pictures by Halbritter, H. In: PalDat - A palynological database.
https://www.paldat.org/pub/Ribes_petraeum/304865; accessed 2022-03-31 Polar and Equatorial

Red Currant

Family :	Grossulariaceae (Currants)	Latin Name :	*Ribes rubra*
Common Name :	Red Currant	Alternate Names :	None
Description :	A shrub normally growing to 1-1.5 m	Flowering times :	April - May

Flowers : the flowers are inconspicuous yellow-green, in pendulous 4-8 cm drooping racemes.

Leaves : five lobed leaves arranged spirally around the stem.

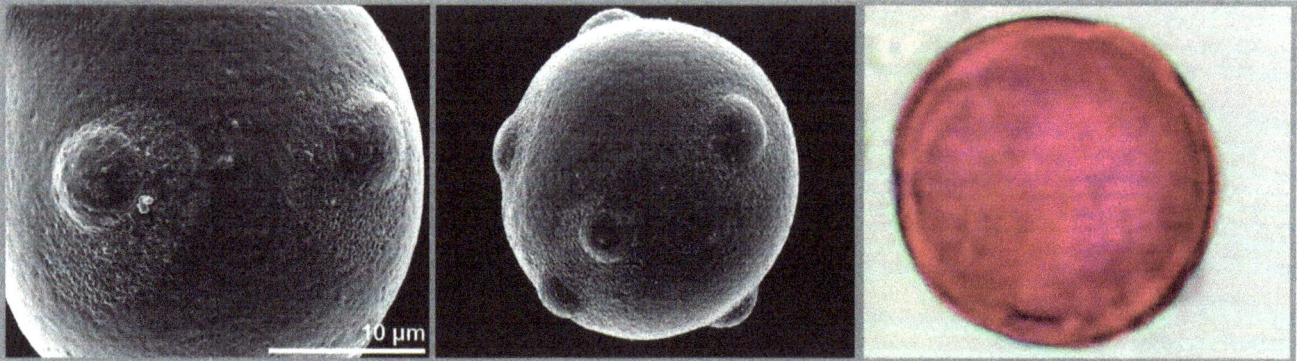

Pollen : Pollen loads are yellow. The pollen grains are 30 um in diameter. There are multiple apertures, in this case, pores and the exine appears of medium thickness. The surface structure of the exine is smooth.

Nectar presentation : No honey crop.

Preferred habitat : Wet woodland fens or shaded banks of streams.

Particular special features : Unlike black currants the leaves, when crushed, do not smell of the fruit.

SEM pictures by Halbritter, H. In: PalDat - A palynological database.
https://www.paldat.org/pub/Ribes_sanguineum/304493; accessed 2022-03-31 Polar and Pores

Gooseberry

Family : Grossulariaceae (Currant Family)
Common Name : Gooseberry

Latin Name : *Ribes uva-crispa*
Alternate Names : None

Description : Bush producing an edible fruit

Flowering times : March - May

Flowers : bell-shaped flowers are produced, singly or in pairs.

Leaves : groups of rounded, deeply crenated 3 or 5 lobed leaves.

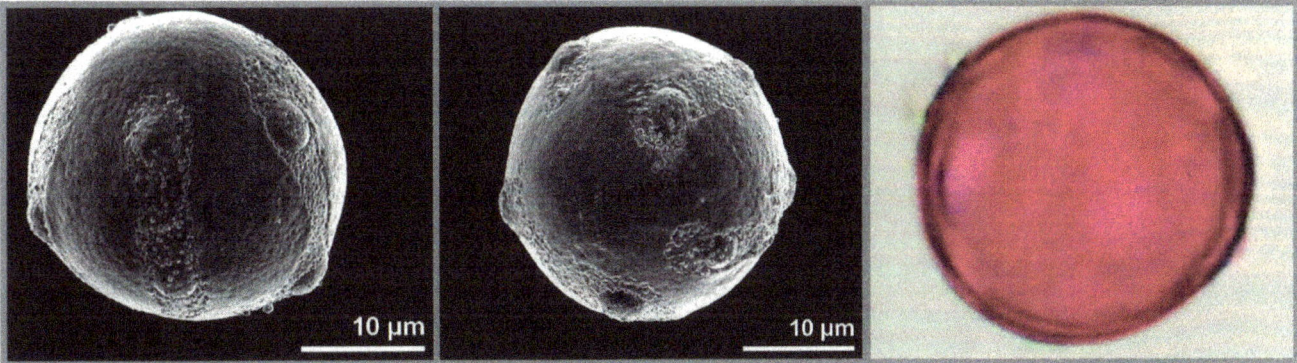

Pollen : Pollen loads are yellow. The pollen grains are 30 um in diameter. There are multiple apertures, in this case, pores and the exine appears of medium thickness. The surface structure of the exine is smooth.

Nectar presentation : No honey crop.

Preferred habitat : Hedgerows, woodland, shaded banks of streams. They prefer fertile, well drained, moisture retentive soil.

Particular special features : The gooseberry is an early flowering fuit and as such provides good early forage.

SEM pictures by Halbritter, H. In: PalDat - A palynological database.
https://www.paldat.org/pub/Ribes_uva-crispa/304002; accessed 2022-03-31 Polar and Equatorial

Fabaceae - Pea Family or Legumes

Key words : Banner, wings and keel. Pea-like pods and often pinnate leaves

The Fabaceae family, is a large and economically important family of flowering plants.

Easy to recognise from the flower as they have 5 united petals that form a 'Banner, wings and keel'.

The banner is a single petal with 2 lobes. There are 2 petals that are the wings, one either side. Then the last 2 petals form the keel and are often fused together.

There are usually 10 stamens but sometimes only 5. There is a partially inferior ovary consisting of a single carpel. This matures into a pea-like pod.

The leaves are very often pinnate.

Lamiaceae are known to sometimes contain pyrrolizidine alkaloids (PAs) - see page 112

Species of importance to bees :

1. Common Name : Broad Bean
   Latin Name : *Vicia faba*

2. Common Name : Red Clover
   Latin Name : *Trifolium pratense*

3. Common Name : White Clover
   Latin Name : *Trifolium repens*

4. Common Name : Sainfoin
   Latin Name : *Onobrychis viciifolia*

5. Common Name : Bird's-foot trefoil
   Latin Name : *Lotus corniculatus*

6. Common Name : Gorse
   Latin Name : *Ulex europaeus*

7. Common Name : Broom
   Latin Name : *Cytisus scoparius*

Red flour beetle on red clover

Broad Bean

Family : Fabaceae
Common Name : Broad Bean

Latin Name : *Vicia faba*
Alternate Names : Field bean

Description : It is a stiffly erect plant 0.5-1.8 m tall
Flowering times : May - July

Flowers : the flowers are 1-2.5 cm long, with five petals, the standard petal white, the wing petals white with a black spot.

Leaves : the leaves are 10-25 cm long, pinnate with 2-7 leaflets, and of a distinct glaucous grey-green colour.

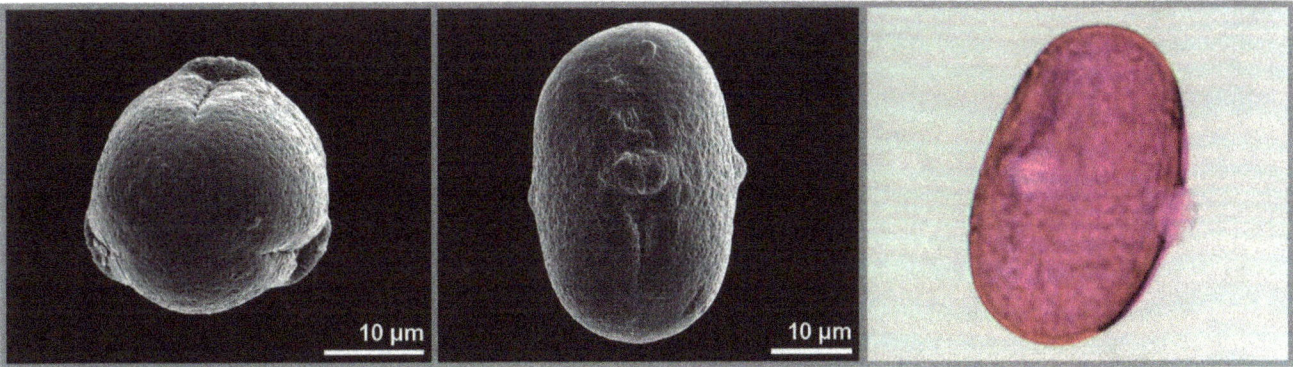

Pollen : Pollen loads are grey / green. The pollen grains are 40 x 20 um. The shape is oblongate. There are 3 apertures which are furrows with pores. The surface structure is netted or pitted and the exine is thin.

Nectar presentation : Major honey crop. Mild flavour. Light colour. Rapid course granulation

Preferred habitat : Field grown and either spring or autumn sown.

Particular special features : Nectar is produced in the deep flowers and only long-tongued bees can access it from the front. Some short-tongued bees are able to bite a hole in the base of the flower to access the nectar. Honey bees will then find these holes and collect the nectar. In addition, field beans have extra-floral nectaries on the underside of the stipules (small leaf-like structures). These nectaries secrete nectar in sunny weather.

When a long-tongued bee lands on the flower the lower petals are pushed down and the bee is dusted with pollen on its underside and taking some of the pollen from the bee. The flower is said to have been 'tripped'.

The broad bean is subject to infestation by aphids and can be a source of 'honeydew'. This is the black bean aphid or *Aphis fabae*.

It has a Pollen Coefficient of 35 which means it is 'normally' represented in honey.

SEM pictures by Halbritter, H. In: PalDat - A palynological database.
https://www.paldat.org/pub/Vicia_cracca/306210; accessed 2022-03-31 Polar and Equatorial

Red Clover

Family :	Fabaceae (Currants)	Latin Name :	*Trifolium pratense*
Common Name :	Red Clover	Alternate Names :	None
Description :	A herbaceous, short-lived perennial	Flowering times :	May - September

Flowers : the flowers are light pink with a yellowish base, 10-15 mm long, produced in a dense inflorescence.

Leaves : the leaves are alternate, trifoliate (with three leaflets), each leaflet 15-30 mm long and 8-15 mm broad.

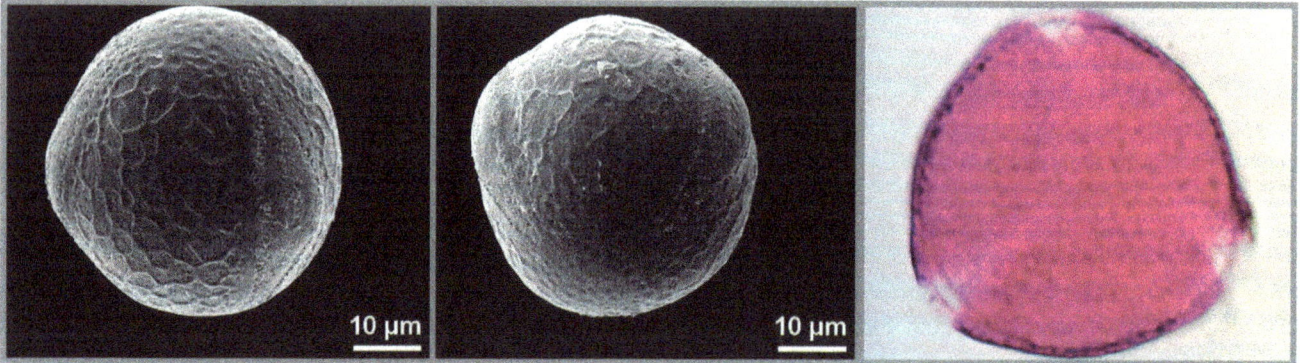

Pollen : Pollen loads are brown. The pollen grains are 30 um in diameter. There are three apertures which are furrows with pores. The surface structre is netted. The exine is thin. The aperture suface is granulated.

Nectar presentation : Medium honey crop. Mild flavour. Light amber colour. Rapid granulation.

Preferred habitat : Pastures, meadows, rough grassland and verges.

Particular special features : The depth of the corolla tube of the flower is approximately 9.5 mm. This restricts access to the nectar to long-tongued bees. However, if the clover is not worked by the longer-tongued bees then it is possible that the honey bees can access the nectar as it builds up and moves up the corolla by capillary action. When the honey bees can reach the nectar they are able to empty the flowers and as such red clover is a significant source of nectar for honey bees. For both red and white clover, the individual florets are held errect until pollination occurs and then they wilt or droop. This tells visiting insects not to bother with that prticular floret and the insects go to the errect florets in preference.

It has a Pollen Coefficient of 25 which means it is 'normally' represented in honey.
The pollen coefficient is a measure of how much pollen is naturally found in the nectar of a given plant. This is largely as a result of the structure and nature of the plant. For example if the anthers are directly above the nectaries and the flower is open and upward facing then it is logical that there may be more pollen found in the nectar than that of a flower with a different structure. The pollen coefficient is used to calculate the amount of pollen in a sample of honey, taking into account the nature of the plant. Using this calculation you can then determine whether the relative amounts of pollen mean that the honey is mono-floral or if it is a true mixture and so multi-floral.

SEM pictures by Halbritter, H. In: PalDat - A palynological database.
https://www.paldat.org/pub/Trifolium_pratense/306432; accessed 2022-03-31 Polar and Equatorial

White Clover

Family :	Fabaceae	Latin Name :	*Trifolium repens*
Common Name :	White Clover	Alternate Names :	None
Description :	Herbaceous perennial plant	Flowering times :	June - September

Flowers : heads of whitish flowers, often with a tinge of pink or cream that may come on with the aging of the plant. The heads are generally 1.5 - 2 cm wide.

Leaves : the leaves are trifoliolate, smooth, elliptic to egg-shaped and long-petioled and usually with light or dark markings.

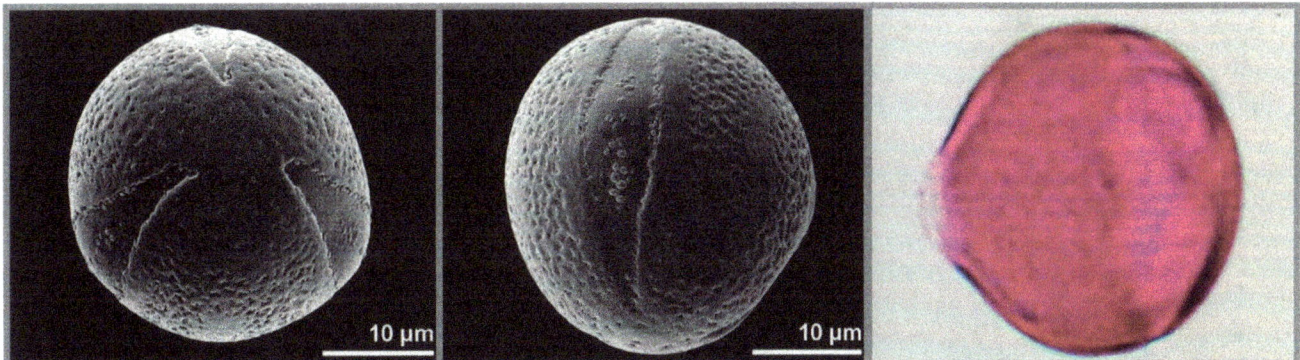

Pollen : Pollen loads are dull green. The pollen grains are 30 um in diameter. There are 3 apertures which are furrows with pores. The surface structure is netted and the exine is thin. The surface of the apertures is granulated.

Nectar presentation : Major honey crop. Mild flavour with delicate after-taste. Light colour. Slow fine granulation. The sugar content of the nectar is circa 40%.

Preferred habitat : Meadows, pastures calcareous grasslands and lawns.

Particular special features : It has a shorter corolla than red clover and so this enables the bumblebees, solitary bees and honey bees all to get the nectar. It used to account for circa 75 % of the British honey crop when clover was used to fortify grazing pasture. Nitrogenous fertilisers and the reduction of pasture has led to a reduction of the abundance of white clover. It is also said that the modern varieties are not as prolific as the old varieties and so are not as attractive to bees as they once were.

The clover head is made up of between 50 and 100 florets. They are initially erect and white but as they get pollinated they wither and change colour. This tells the bees not to visit this floret and makes for efficient pollination. The clover flowers yield nectar at a relatively low temperature but stop producing nectar when the temperature reaches 25 degrees Celsius.

It has a Pollen Coefficient of 50 which means it is 'normally' represented in honey.

SEM pictures by Halbritter, H. In: PalDat - A palynological database.
https://www.paldat.org/pub/Trifolium_repens/306197; accessed 2022-03-31 Polar and Equatorial

Sainfoin

Family :	Fabaceae (Currants)	Latin Name :	*Onobrychis viciifolia*
Common Name :	Sainfoin	Alternate Names :	None
Description :	A deep-rooted perennial legume	Flowering times :	June - August

Flowers : showy and pink, white or purple and tightly arranged in a compact raceme with 20 to 50 flowers per head. A stalked spike of bright pink flowers.

Leaves : odd-pinnately compound with 11 to 21 leaflets.

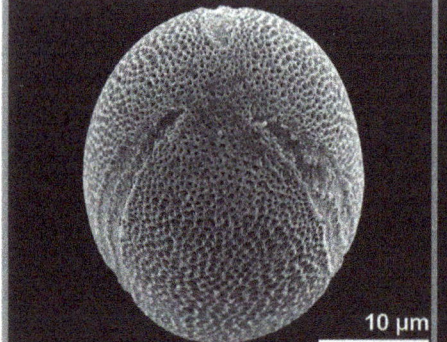
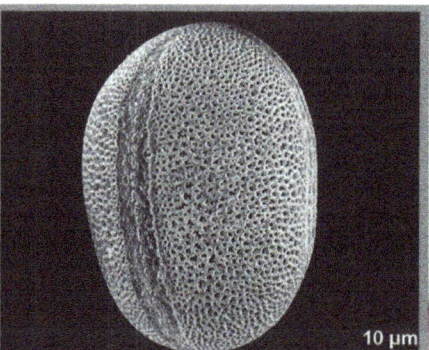
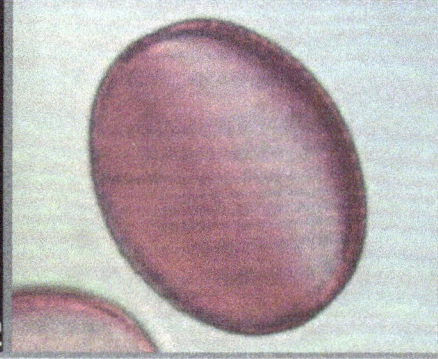

Pollen : Pollen loads are dark brown. The pollen grains are 20 x 40 um in size and they are long in shape. There are 3 apertures which are furrows. The surface structure is netted or pitted with the exine appearing thin. The structure of the apertures has a cap or a streak.

Nectar presentation : Medium honey crop. Sweet pronounced flavour. Pale yellow colour. Rapid course granulation. Odour is intially poor but this disappears over time. The resulting wax has a lovely pale yellow colour which is also transferred to the frames. The colour comes from the oils in the pollen.

Preferred habitat : Unimproved chalk grassland, tracksides and road verges.

Particular special features : It has a deep penetrating tap root which renders it very insensitive to surface moisture. This makes it a reliable nectar producer. The flowers are good nectar producers and are considerd one of the best nectar plants. They will even secrete nectar at temperatures as low as 14 degrees Celsuis.

It has a Pollen Coefficient of 75 which means it is 'normally' represented in honey.

SEM pictures by Halbritter, H. In: PalDat - A palynological database.
https://www.paldat.org/pub/Onobrychis_viciifolia/306131; accessed 2022-03-31 Polar and Equatorial

Bird's-foot trefoil

Family :	Fabaceae	Latin Name :	*Lotus corniculatus*
Common Name :	Bird's-foot trefoil	Alternate Names :	'Granny's Toenails', 'Butter and Eggs'
Description :	A perennial herbaceous plant	Flowering times :	June - September

Flowers : yellow, 10 - 16 mm long. 5 petals; the lateral two the 'wings', the lower two united to form the 'keel', overall shape of corolla being butterfly-like.

Leaves : five leaflets are present, but with the central three held conspicuously above the others, hence the use of the name 'trefoil'.

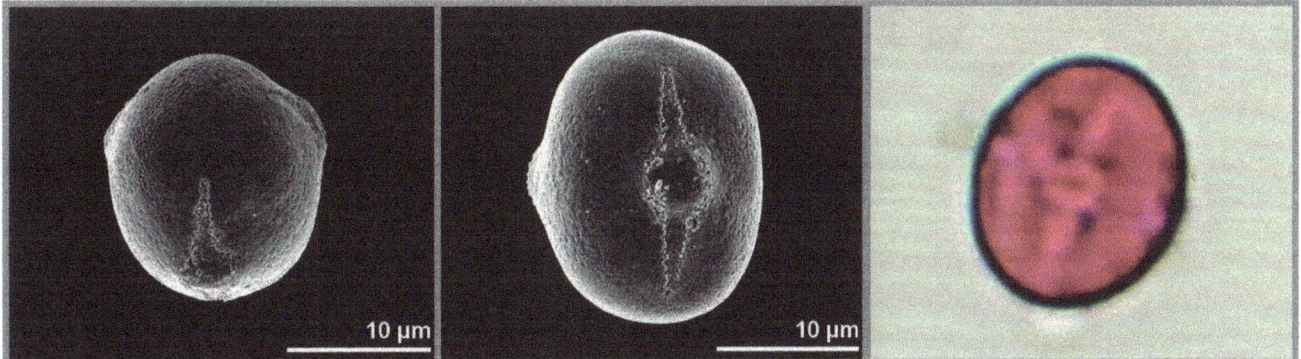

Pollen : Pollen loads are light brown. The pollen grains are 15 x 20 um. There grains are oval elongated in shape. There are 3 apertures which are furrows with pores. The surface structure is netted or pitted and the exine appears to be thin. There are no apparent surface features on the apertures.

Nectar presentation : Not a significant honey crop in the UK - Honey is light and tastes like clover. Rapid granulation.

Preferred habitat : Calcareous grassland, meadows, hill pasture, grass heaths, shingle, cliffs and sand dunes.

Particular special features : Like Sainfoin, it is insensitive to surface moisture due to a large tap root. Hence it is a good nectar source for a range of bee species including honey bees, short and long-tongued bumblebees and solitary bees.

It has a Pollen Coefficient of 25 which means it is 'normally' represented in honey.

SEM pictures by Halbritter, H. In: PalDat - A palynological database.
https://www.paldat.org/pub/Lotus_corniculatus/306119; accessed 2022-03-31 Polar and Equatorial

Gorse

Family : Fabaceae (Currants)
Common Name : Gorse
Latin Name : *Ulex europaeus*
Alternate Names : Furze or Whin

Description : Species of thorny evergreen shrubs
Flowering times : December - June

Flowers : yellow flowers, generally showy, some with a very long flowering season. Gorse flowers have a distinctive coconut scent

Leaves : the leaves of young plants are trifoliate but in mature plants they are reduced to scales or small spines.

Pollen : Pollen loads are orange - brown. The pollen grains are 40 um in diameter. The grains are oval but triangular in cross-section. There are 3 apertures which are furrows. The surface structure is smooth and the exine appears thin.

Nectar presentation : No honey crop.

Preferred habitat : Heathlands, commons, road-sides, sea cliffs, waste ground on well drained acid soils or coastal calcareous soils.

Particular special features : A good source of early pollen for a number of bee species including honey bees, bumblebees and solitary bees.

Broom

Family :	Fabaceae	Latin Name :	*Cytisus scoparius*
Common Name :	Broom	Alternate Names :	Common broom or Scotch broom
Description :	A perennial leguminous shrub	Flowering times :	May - June

Flowers : in spring and summer is covered in profuse golden yellow flowers 20-30 mm from top to bottom and 15-20 mm wide.

Leaves : the shrubs have green shoots with small deciduous trifoliate leaves 5-15 mm long

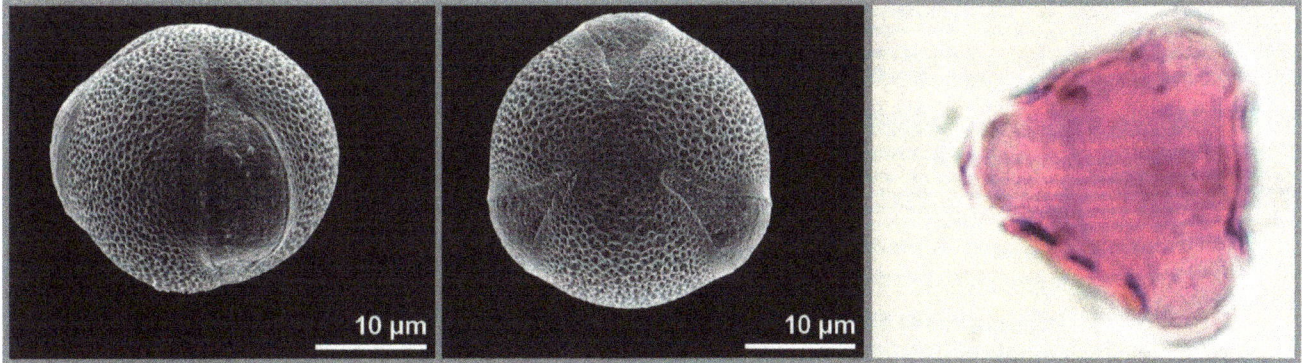

Pollen : Pollen loads are orange. The pollen grains are 25 um in diameter. The grains are oval but triangular in cross-section. The surface structure is netted or pitted. There are 3 apertures which are furrows with pores. There are no apparent surface features on the apertures.

Nectar presentation : No honey crop.

Preferred habitat : Heathlands, open woodland, roadside verges and railway embankments on light sandy soils.

Particular special features : There is some debate whether broom provides any nectar, however, it is a good supply of pollen for a range of bee species including honey bees, bumblebees and solitary bees.

Note - for both gorse and broom, there are more than one species of both genera - especially when you take into account garden varieties.

SEM pictures by Halbritter, H. In: PalDat - A palynological database.
https://www.paldat.org/pub/Cytisus_nigricans/305247; accessed 2022-03-31 Polar and Equatorial

Rosaceae - Rose Family

Key words : 5 sepals and 5 petals with numerous stamens. Oval serrated leaves

A medium-sized family of flowering plants, including 4,828 known species in 91 genera.

The name is derived from the type genus Rosa. Among the most species-rich genera are Alchemilla (270), Sorbus (260), Crataegus (260), Cotoneaster (260), Rubus (250), and Prunus with about 200 species. However, all of these numbers should be seen as estimates - much taxonomic work remains to be done.
The family Rosaceae includes herbs, shrubs, and trees. Most species are deciduous, but some are evergreen. They have a worldwide range, but are most diverse in the Northern Hemisphere.

The rosaceae have alternate leaves which can be simple, trifoliate, palmate or pinnate. They are usually (more or less) oval shaped with serrated edges. Members of the rose family have an epicalyx and sepals - if these are present then it has to be a rose.

The flowers typically have 5 separate sepals and a similar number of petals.
There are a minimum of 5 stamens but very often many more
They tend to have many simple pistils resulting in a fuzzy-looking centre surrounded by lots of stamens

The rose family produces many edible fruits.

Species of importance to bees :

1. Common Name : Hawthorn
    Latin Name :   **Crataegus spp**

2. Common Name : Cherry
    Latin Name :   **Prunus spp**

3. Common Name : Apple
    Latin Name :   **Malus pumila**

4. Common Name : Blackberry
    Latin Name :   **Rubus fruticosus**

5. Common Name : Cherry Laurel
    Latin Name :   **Prunus laurocerasus**

6. Common Name : Cotoneaster
    Latin Name :   **Cotoneaster spp**

7. Common Name : Raspberry
    Latin Name :   **Rubus idaeus**

8. Common Name : Pear
    Latin Name :   **Pyrus spp**

9. Common Name : Rowan
    Latin Name :   **Sorbus spp**

10. Common Name : Strawberry
    Latin Name :   **Fragaria spp**

11. Common Name : Blackthorn
    Latin Name :   **Prunus spinosa**

Tawny mining bee *Andrena fulva* on apple flowers in April 2020

Buff-tailed Bumblebee *Bombus terrestris* foraging on apple flowers in April 2020

Early Bumblebee *Bombus pratorum* foraging on raspberry flowers in May 2020

Honey bee foraging on pear flowers in March 2020

Hawthorn

Family : Rosaceae
Common Name : Hawthorn
Description : A large genus of shrubs and trees

Latin Name : *Crataegus* spp
Alternate Names : Thornapple, May-tree or Hawberry
Flowering times : March - June

Flowers : perfect, usually small white flowers, with 5 petals produced in clusters near the end of the twig

Leaves : highly variable, but generally alternate, simple, 50 to 100 mm long, serrated and lobed (may be unlobed), long thorns, dark green above and paler below.

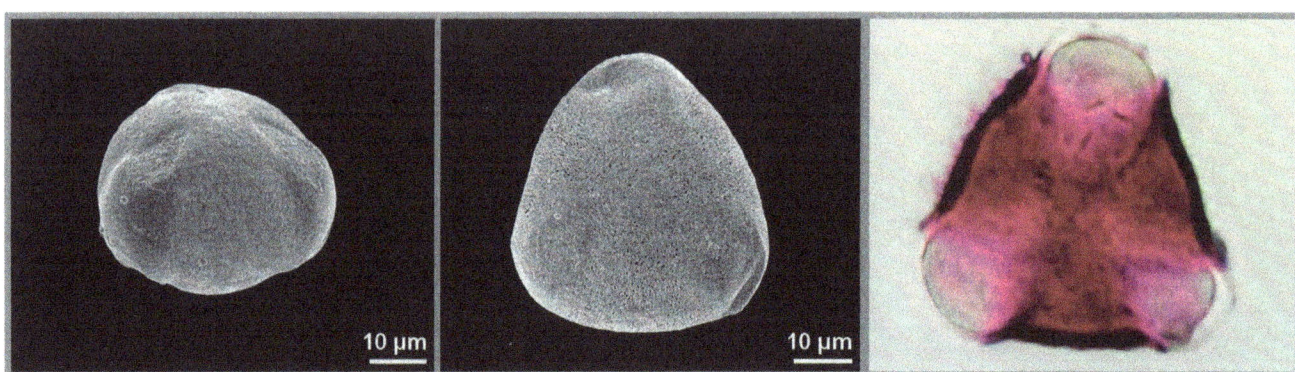

Pollen : Pollen loads are green. The pollen grains are 40 um in diameter. The shape of the pollen grains is oval but triangular in cross-section. There are 3 apertures being furrows with pores. The surface structure is smooth or indefinite and there are no apparent surface features on the apertures.

Nectar presentation : No honey crop - exceptionally rare producing a dark amber honey with a rich flavour. Sometimes with a greenish tinge.

Preferred habitat : Hedges, scrub and open woodland in all soils except acid peat.

Particular special features : The flowers are shallow which allows even short-tongued bees to access the nectar. Therefore, it is visited by a wide range of bee species including honey bees. The hawthorn is said to yield a good honey crop once in 7 years and no-one seems to know why. It does not seem to be related to weather or soil conditions. Some sources say that the air temperature needs to be consistently above 25 degrees Celsius to yield nectar in any quantity.

SEM pictures by Bombosi, P. In: PalDat - A palynological database.
https://www.paldat.org/pub/Crataegus_monogyna/304040; accessed 2022-03-31 Polar and Equatorial

Cherry

Family : Rosaceae
Common Name : Cherry

Latin Name : ***Prunus* spp**
Alternate Names : Stone fruit

Description : A genus of trees and shrubs

Flowering times : March - April

Flowers : are usually white to pink, sometimes red, with five petals and five sepals. There are numerous stamens. Flowers are borne singly, or in umbels of two to six or sometimes more on racemes.

Leaves : simple, alternate, usually lanceolate, unlobed, and often with nectaries on the leaf stalk.

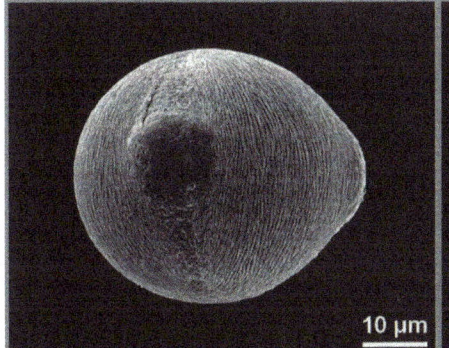
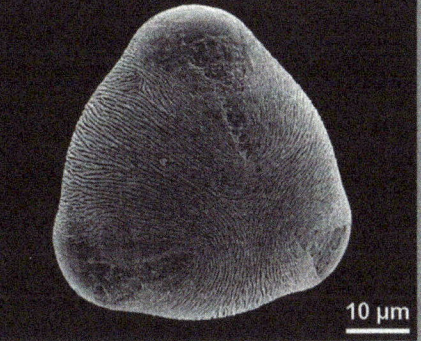

Pollen : Pollen loads are orange - brown. The pollen grains are 35 um in diameter. The shape of the pollen grains is oval but triangular in cross-section. There are 3 apertures being furrows with pores. The surface structure is striated and there are no apparent surface features on the apertures.

Nectar presentation : Minor honey crop. Pale yellow colour. Granulation rapid

Preferred habitat : Woods, commons, hedgerows and verges.

Particular special features : Cherry orchards are a common polination contract for beekeepers. Needing the honey bees to be in the orchards during blossoming time as the numbers of wild bees, this early in the season, are low. A honey crop from the cherries is possible given good weather but this seems to be rare. The cherry flowers allow access for a range of bee species to access the nectaries.

It has a Pollen Coefficient of 25 which means it is 'normally' represented in honey.

SEM pictures by Halbritter, H. In: PalDat - A palynological database.
https://www.paldat.org/pub/Prunus_dulcis/301737; accessed 2022-03-31 Polar and Equatorial

Apple

Family :	Rosaceae	Latin Name :	*Malus pumila*
Common Name :	Apple	Alternate Names :	None
Description :	A deciduous tree in the rose family	Flowering times :	April

Flowers : the 3 to 4 cm flowers are white with a pink tinge that gradually fades, five petaled, with an inflorescence consisting of a cyme with 4-6 flowers.

Leaves : are alternately arranged dark green-coloured simple ovals with serrated margins and slightly downy undersides.

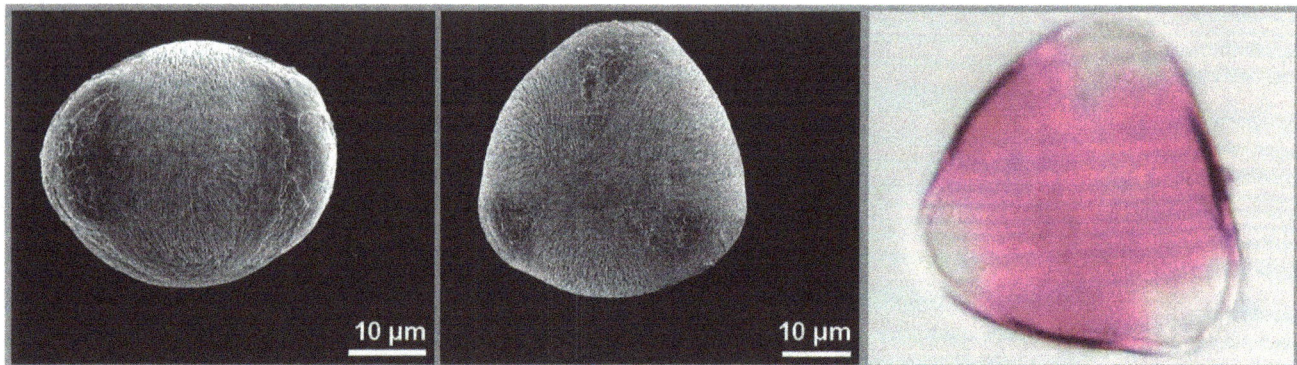

Pollen : Pollen loads are yellow-brown. The pollen grains are 40 um in diameter. The shape of the pollen grains is oval but triangular in cross-section. There are 3 apertures being furrows with pores. The surface structure is striated and there are no apparent surface features on the apertures.

Nectar presentation : Minor honey crop. Amber colour with aroma of apples. Granulation irregular.

Preferred habitat : Woodland, scrub and hedgerows on fertile, humous rich, soils.

Particular special features : The open flowers allow access to a wide range of bee species. Honey bees are brought to apple orchards in pollination contracts. As apple is the last of the fruiting trees (of value to bees) to flower it is more likely to offer a honey crop. However, the value of the apple blossom is really in being an ealry nectar source on which the honey bees can raise brood. Another advantage of apple blossom is that the different varieties of apple bloom at slightly different times. Each tree may only be blossoming for 2 weeks but if there are several varieties then there may be blossom around for far longer.

The apple flowers are open and as such may have the nectar diluted by rain or dew. This will sometimes reduce the foraging of honey bees for nectar and turn them to foraging for apple pollen instead. The honey bees returning to foraging for nectar once the rain or dew has evaporated off.

It has a Pollen Coefficient of 25 which means it is 'normally' represented in honey.

SEM pictures by Halbritter, H. In: PalDat - A palynological database.
https://www.paldat.org/pub/Malus_sylvestris/306281; accessed 2022-03-31 Polar and Equatorial

Blackberry

Family : Rosaceae
Common Name : Blackberry

Latin Name : *Rubus fruticosus*
Alternate Names : Bramble

Description : An edible fruit producing shrub

Flowering times : May - September

Flowers : are produced in late spring and early summer on short racemes on the tips of the flowering laterals. Each flower is about 2-3 cm in diameter with five white or pale pink petals.

Leaves : large palmately compound leaves with five or seven leaflets.

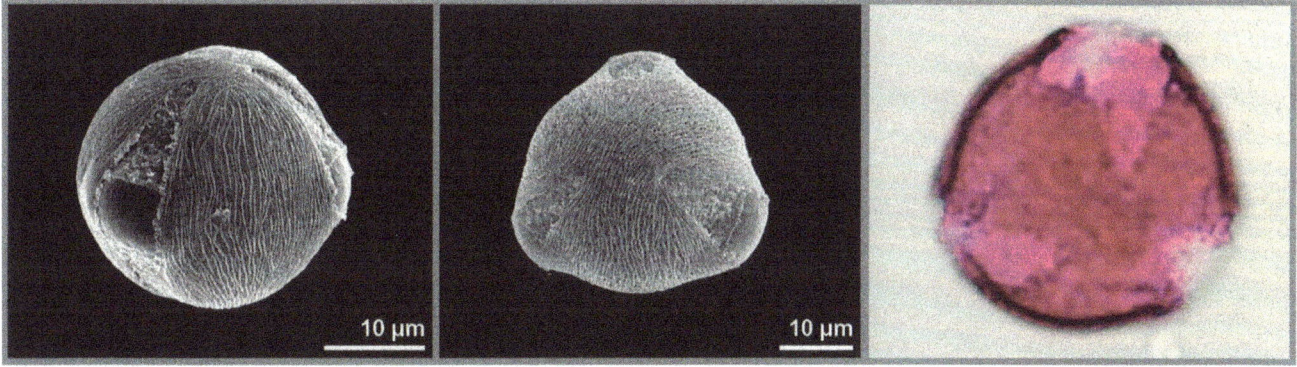

Pollen : Pollen loads are dull green. The pollen grains are 30 um in diameter. The shape of the pollen grains is oval but triangular in cross-section. There are 3 apertures which are forrows with pores. The surface structure is smooth or indefinite and the exine appears of medium thickness. There are no obvious features on the surface of the apertures.
Nectar presentation : Important honey crop. Light colour with a delicate flavour like clover. Slow granulation.

Preferred habitat : Woodland, hegderows, commons, scrub, heaths and cliffs on light but fertile soils.

Particular special features : There are over 320 microspecies of blackberry. This is a brilliant plant for all bees and other insects such as beetles, butterflies and wasps. The particular character of blackberry that is of importance is the nature of its deep roots. This makes the plant insensetive to surface moisture and it is a reliable necture producing plant when most others have stopped in drought conditions. It has become a very important honey crop for many UK beekeepers in recent years.

One interesting point is that flowers on different parts of the plant will secrete different amounts of nectar. Flowers on thick shoots produce more nectar than those on thin shoots.

It has a Pollen Coefficient of 50 which means it is 'normally' represented in honey.

Note - For both apple and blackberry there are many species and sub-species so it might have been better to have written *Malus* **spp** and *Rubus* **spp.**

SEM pictures by Bombosi, P. In: PalDat - A palynological database.
https://www.paldat.org/pub/Rubus_fructicosus/301336; accessed 2022-03-31 Polar and Equatorial

Cherry Laurel

Family :	Rosaceae	Latin Name :	***Prunus laurocerasus***
Common Name :	Cherry Laurel	Alternate Names :	Common laurel and English laurel
Description :	An evergreen species of cherry	Flowering times :	April - June

Flowers : erect 7-15 cm racemes of 30-40 flowers, each flower 1 cm across, with five creamy-white petals and numerous yellowish stamens with a sweet smell.

Leaves : dark green, leathery, shiny, 10-25cm long and 4-10 cm broad, with a finely serrated margin. The leaves can have the scent of almonds when crushed.

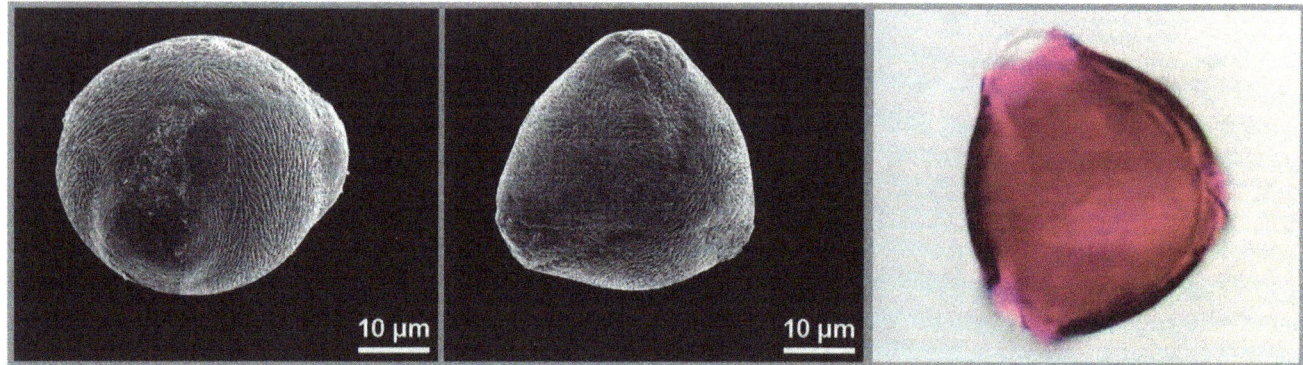

Pollen : Pollen loads are dull green. The pollen grains are 40 um in diameter. The pollen grains are oval but triangular in cross-section. There are 3 apertures which are furrows with pores. The exine appears of medium thickness. The surface structure is smooth or indefinite and there appear to be no suface features on the apertures.

Nectar presentation : No honey crop.

Preferred habitat : Woods and scrubland.

Particular special features : Cherry laurel has extra floral nectaries on the undersides of the leaves. In this way it is able to supply nectar to solitary bees, long and short-tongued bumblebees and honey bees.

SEM pictures by Halbritter, H. In: PalDat - A palynological database.
https://www.paldat.org/pub/Prunus_laurocerasus/306335; accessed 2022-03-31 Polar and Equatorial

Cotoneaster

Family : Rosaceae
Common Name : Cotoneaster

Latin Name : *Cotoneaster* **spp**
Alternate Names : None

Description : A genus of flowering plants

Flowering times : May - June

Flowers : solitary or in corymbs of up to 100 together. The flower is either fully open or has its five petals half open 5-10 mm diameter. They may be any shade from white through creamy white to light pink to dark pink to red

Leaves : arranged alternately, 0.5-15 cm long, ovate to lanceolate in shape, entire; both evergreen and deciduous

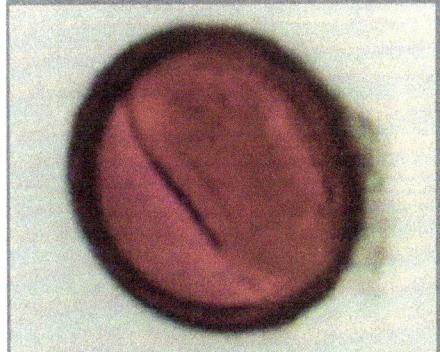
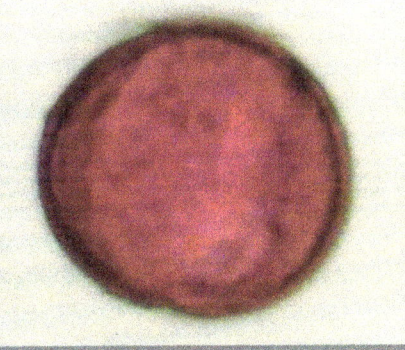
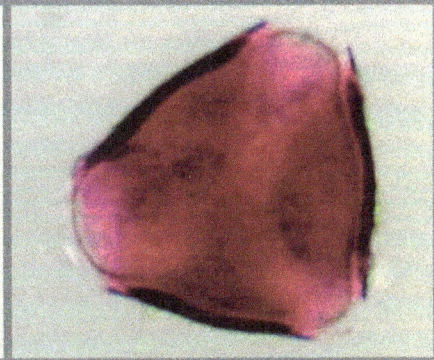

Pollen : Pollen loads are yellow - green. The pollen grains are 30 um in diameter. The pollen grains are oval but triangular in cross-section. There are 3 apertures which are pores. The exine appears to be of medium thickness and the surface structre is smooth or indefinite. There are no surface features on the apertures.

Nectar presentation : No honey crop.

Preferred habitat : Cliffs, walls, roadsides, railway embankments, quarries and rocky grassland.

Particular special features : They produce copious amounts of nectar and attract a wide variety of bee species. Interestingly, they are originally from the Himalayas and as such are classed as non-native invasive species. This prohibits them from being planted or distributed to the wild.

Raspberry

Family : Rosaceae
Common Name : Raspberry
Description : Generally perennials - biennial stems

Latin Name : **Rubus idaeus**
Alternate Name : None
Flowering times : May - August

Flowers : short racemes on the tips of these side shoots, each flower about 1 cm diameter with five white petals.

Leaves : large pinnately compound leaves with five or seven leaflets. In its second year a stem does not grow taller, but produces several side shoots, which bear smaller leaves with three or five leaflets.

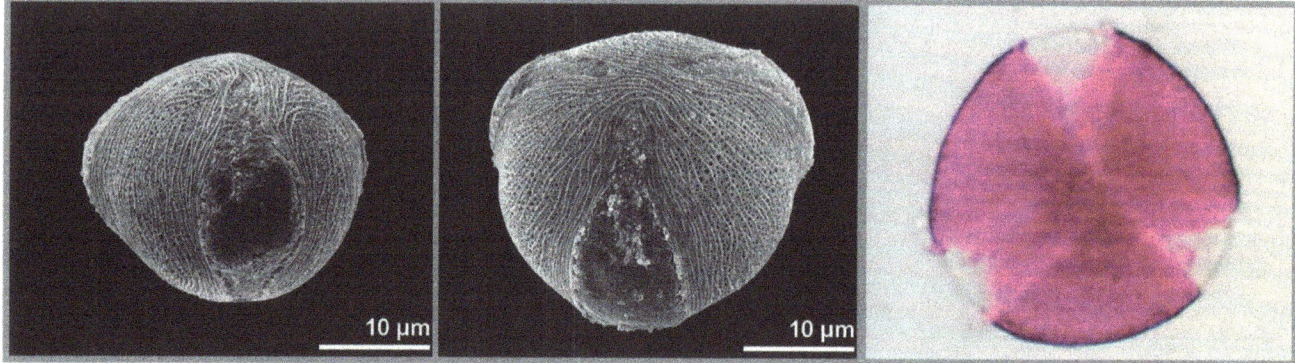

Pollen : Pollen loads are light green. The pollen grains are 25 um in diameter. The pollen grains are oval but triangular in cross-section. The exine appears thin. There are 3 apertures which are furrows with pores. The surface structure is striated and there are no features on the surface of the pores.

Nectar presentation : A minor honey crop. Flavour and aroma is mild with rapid granulation.

Preferred habitat : Open woodland, heath and commons.

Particular special features : Bees are the main pollinators of raspberries and are directly related to the quality and quantity of the fruit produced. Bumblebees being more effective pollinators of raspberries than honey bees. This is probably due to the bumblebees using a 'buzz-pollination' methodology.

The pendulus nature of the raspberry flowers means that the nectar is protected during rain. This means that bees are able to work these flowers when most other flowers have been spoilt by the rain. Raspberry flowers are very attractive to a wide range of bees including short and long-tongued bumblebees, solitary bees and honey bees

It is said that bees will forage the juice from over-ripe fruit and it could be the source of the so called 'red honey' that occurs sometimes.

It has a Pollen Coefficient of 50 which means it is 'normally' represented in honey.

SEM pictures by Halbritter, H. In: PalDat - A palynological database.
https://www.paldat.org/pub/Rubus_caesius/306374; accessed 2022-03-31 Polar and Equatorial

Pear

Family : Rosaceae
Common Name : Pear

Latin Name : *Pyrus* spp
Alternate Names : None

Description : Fruit bearing tree to 17 metres tall

Flowering times : March - April

Flowers : white, rarely tinted yellow or pink, 2-4 centimetres diameter, and have five petals.

Leaves : simple, 2-12 centimetres long, glossy green on some species, densely silvery-hairy in some others; leaf shape varies from broad oval to narrow lanceolate.

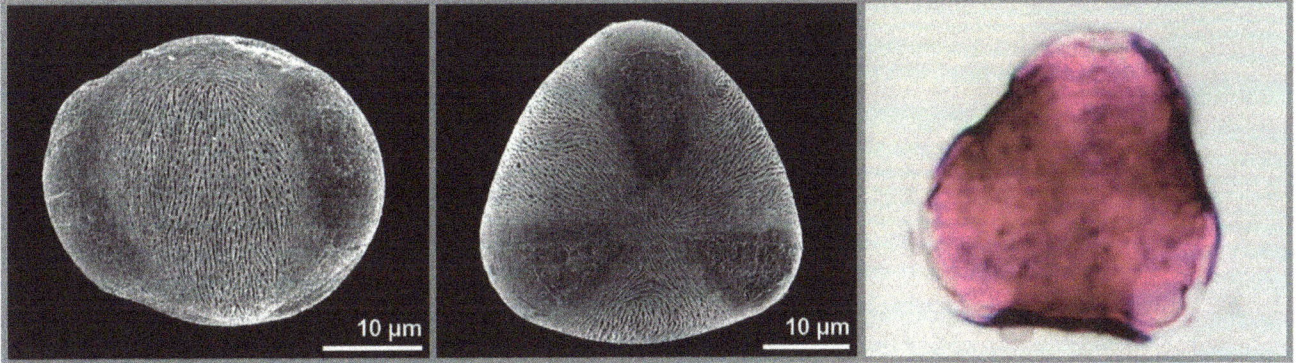

Pollen : Pollen loads are light green. The pollen grains are 35 um in diameter. The pollen grains are oval but triangular in cross-section. There are 3 apertures which are furrows with pores. The surface structure is smooth or indefinite with the exine being of medium thickness. There are no surface features on the apertures.

Nectar presentation : No honey crop.

Preferred habitat : Hedges, wood margins and waste ground.

Particular special features : It is a valuable source of early nectar and pollen in the UK. They flower ahead of apples, however, the different varieties may overlap. The nectar is said to be weaker than that of apple (lower sugar content), however, the pollen is very attractive to bees including solitary bees, short and long-tongued bumblebees and honey bees.

It has a Pollen Coefficient of 25 which means it is 'normally' represented in honey.

SEM pictures by Halbritter, H. In: PalDat - A palynological database.
https://www.paldat.org/pub/Pyrus_pyraster/305137; accessed 2022-03-31 Polar and Equatorial

Rowan

Family : Rosaceae
Common Name : Rowan

Latin Name : *Sorbus* spp
Alternate Name : Whitebeam, Wild service tree, Mountain-a

Description : Small deciduous trees 10-20 m tall

Flowering times : May - June

Flowers : are borne in dense clusters, each one bearing five creamy white petals

Leaves : pinnate comprising 5-8 pairs of leaflets, plus one 'terminal' leaflet. Each leaflet is long, oval and toothed.

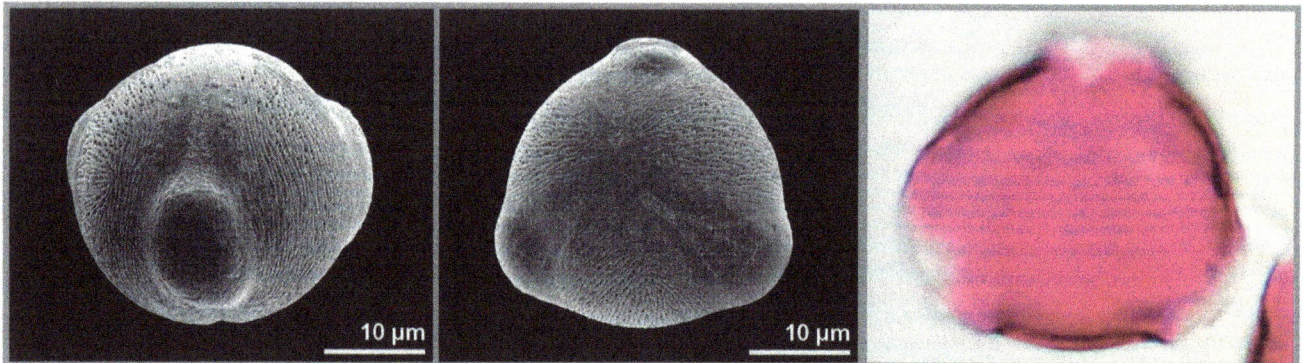

Pollen : Pollen loads are light green. The pollen grains are 30 um in diameter. The pollen grains are oval but triangular in cross-section. There are 3 apertures which are furrows with pores. The exine appears to be of medium thickness and the surface structure is smooth or indefinite. There are no obvious features on the surface of the apertures.

Nectar presentation : No honey crop.

Preferred habitat : Open woodlnad, scrub, mountain rocks, cliffs, rocky river banks on acid soils.

Particular special features : They have very small open flowers that are visited by flies, short-tonged bumblebees and honey bees.

SEM pictures by Bombosi, P. In: PalDat - A palynological database.
https://www.paldat.org/pub/Sorbus_aucuparia/306185; accessed 2022-03-31 Polar and Equatorial

Strawberry

Family : Rosaceae
Common Name : Strawberry

Latin Name : *Fragaria* spp
Alternate Names : None

Description : Herbaceous perennial plant to 15cm

Flowering times : April - July

Flowers : each flower has 5 white petals, 5 green sepals, and 5 green sepal-like bracts. They can be pistillate, staminate, or perfect.

Leaves : consisting of several basal leaves and one or more inflorescences. The basal leaves are trifoliate.

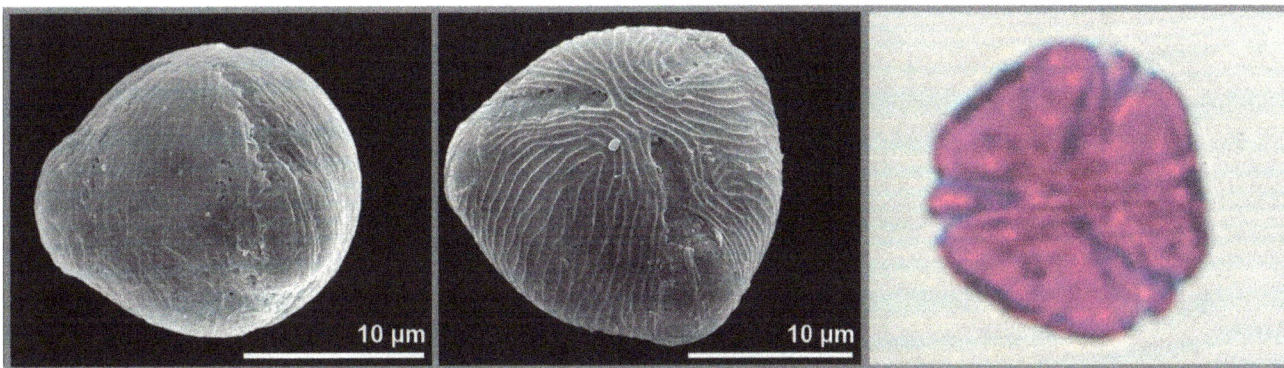

Pollen : Pollen loads are dark green. The pollen grains are 20 um in diameter. The pollen grains are oval but triangular in cross-section. There are 3 apertures which are furrows with pores. The surface structure is smooth or indefinite and the exine appears of medium thickness. There appears to be a cap or streak on the apertures.

Nectar presentation : No honey crop.

Preferred habitat : Woodlands, scrub, hedge banks, rough grassland on rich spoils.

Particular special features : Strawberries are a crop that needs to be insect pollinated for fruit quality and quantity. This is sometimes done by honey bees but is most often done by buff-tailed bumblebees which fly at lower temperatures than honey bees.

SEM pictures by Halbritter, H. In: PalDat - A palynological database.
https://www.paldat.org/pub/Fragaria_viridis/304038; accessed 2022-03-31 Polar and Equatorial

Blackthorn

Family : Rosaceae
Common Name : Blackthorn

Latin Name : **_Prunus spinosa_**
Alternate Name : Sloe

Description : Spiny, densely branched trees to 7m
Flowering times : March - April

Flowers : white flowers appear on short stalks before the leaves in March and April, either singularly or in pairs measuring 5 - 8 cm.

Leaves : slightly wrinkled, oval, toothed, pointed at the tip and tapered at the base.

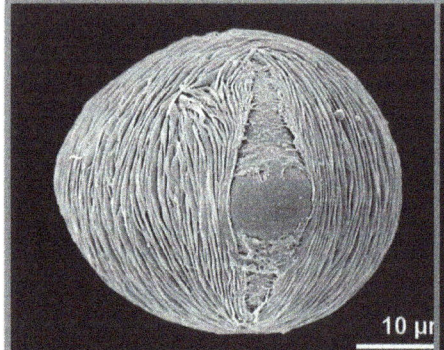
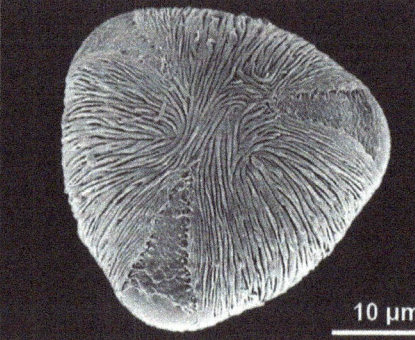
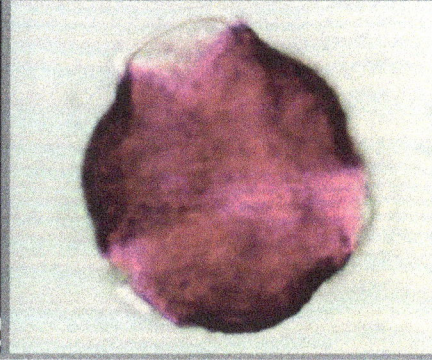

Pollen : Pollen loads are orange. The pollen grains are 40 um in diameter. There pollen grains are oval but triangular in cross-section. There are 3 apertures which are furrows with pores. The surface structure is striated and the exine appears of medium thickness. There does not appear to be any surface features on the apertures.

Nectar presentation : No honey crop.

Preferred habitat : Hedgerows, scrub, wood margins, cliff tops and shingle beaches.

Particular special features : The flowers form before the leaves and offer an early source of nectar and pollen. However, the weather usually prevents honey bees from taking advantage of this. Consequently, the blackthorn is visited by early bumblebees and solitary bees.

SEM pictures by Halbritter, H. In: PalDat - A palynological database.
https://www.paldat.org/pub/Prunus_spinosa/306340; accessed 2022-03-31 Polar and Equatorial

Salicaceae - Willow Family

Key words : Bushes or trees with alternate leaves in moist soils. Catkins form small capsules

Recent genetics have greatly expanded the circumscription of the family to contain 56 genera and about 1220 species. They are a family of bushes and trees with simple alternate leaves. The plants are dioecious - with male and female flowers appearing on separate plants.

The flowers are unisexual with both male and female appearing in catkins.
The sepals are greatly reduced or absent and there are no petals.

Male flowers have 2 or more stamens
Female flowers have a superior ovary consisting of 2 or 4 united carpels as indicated by the stigmas.

Carpels are fused to create a single chamber which matures as a capsule, usually with fine silken hairs to help transport the seeds by air.

In the UK, willows are a really difficult group to identify, with over 70 species in the wild and horticulture.

Species of importance to bees :

1. Common Name : Musk Willow
   Latin Name :   *Salix aegyptiaca*

2. Common Name : Pussy Willow
   Latin Name :   *Salix caprea*

Honey bee foraging on Salix aegyptiaca - photograph courtesy of Teeranlall Ramgopal.

Musk Willow

Family :	Salicaceae	Latin Name :	***Salix aegyptiaca***
Common Name :	Musk Willow	Alternate Name :	Calaf of Persia willow
Description :	Deciduous shrubs and trees	Flowering times :	February - March

Flowers : are dioecious. Catkins are fragrant, grey, males to 30 mm long with yellow anthers, females to 75 mm long.

Leaves : are simple, green, shiny, alternate and are 5 to 15 cm long and 3 to 6 cm wide.

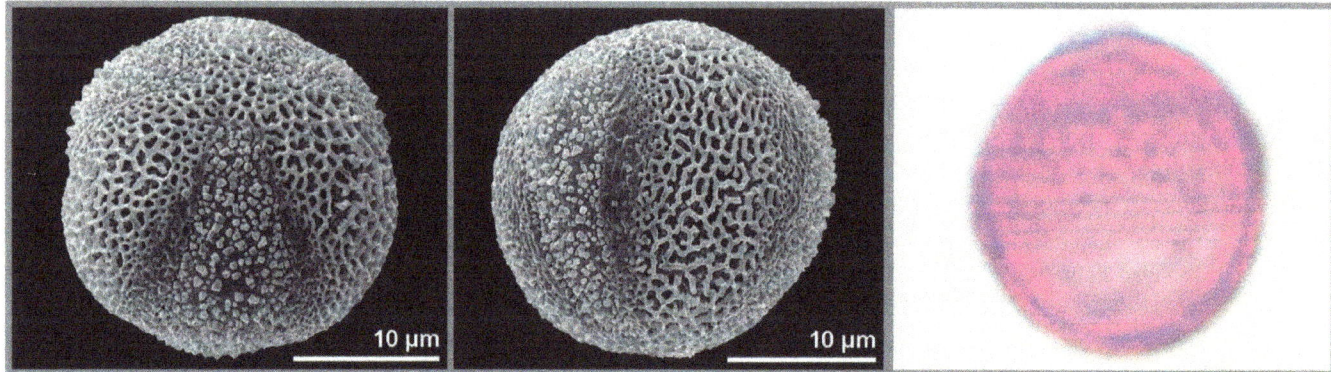

Pollen : Pollen loads are yellow. The pollen grains are 20 um in diameter. The pollen grains are round to oval in shape. There are 3 apertures which are furrows with pores. The surface structure is netted or pitted and the exine appears to be medium in thickness. There are no apparent surface features on the apertures.

Nectar presentation : No honey crop.

Preferred habitat : Open woodland, scrub, hedgerows, waste ground, lake and stream margins being moderately fertile, moist but well drained.

Particular special features : The catkins are present before the leaves appear. The male catkins present the pollen in abundance and it is very attractive to bees.

Aegyptiaca is the earliest flowering willow and the local record is the 20th January for a mature tree. The one in the photograph above is about 10 years old and the first pollen, this year, was offered on the 25th January 2020.

There is nectar presented from both male and female flowers but not in massive quantities.

SEM pictures by Halbritter, H. In: PalDat - A palynological database.
https://www.paldat.org/pub/Salix_fragilis/301226; accessed 2022-03-31 Polar and Equatorial

Pussy Willow

Family : Salicaceae
Common Name : Pussy Willow

Latin Name : *Salix caprea*
Alternate Names : Goat willow, Great sallow

Description : Deciduous shrubs and trees

Flowering times : March - April

Flowers : male catkins are grey, stout and oval, which become yellow when ripe with pollen. Female catkins are longer and green. Pussy willow is dioecious.

Leaves : are oval rather than long and thin. They are hairless above, with a felty coating of fine grey hairs underneath, and a pointed tip which bends to one side.

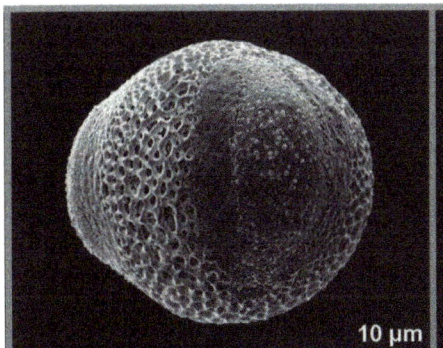

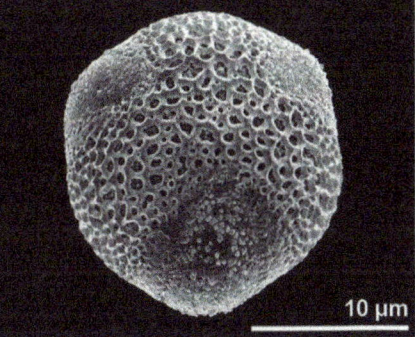

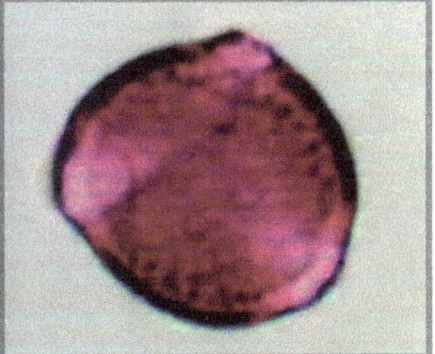

Pollen : Pollen loads are yellow. The pollen grains are 20 um in diameter. The pollen grains are round to oval in shape. There are 3 apertures which are furrows with pores. The surface structure is netted or pitted and the exine appears to be medium in thickness. There are no apparent surface features on the apertures.

Nectar presentation : Not a significant crop. Light golden yellow. Mild flavour. Fine granulation.

Preferred habitat : Open woodland, scrub, hedgerows, waste ground, lake and stream margins being moderately fertile, moist but well drained.

Particular special features : Great necatar and pollen plants for honey bees, short and long-tongued bumblebees and solitary bees too. Male flowers emit a higher total amount of scent (4-5 times that of the female flowers), a phenomenon also seen in other dioecious plants. This is not because of the increased weight of the male catkins but specifically to encourage greater foraging. The substances emitted in the scents are the same for male and female flowers.

Salix cuttings are an easy way to produce more plants. Take a pencil sized cutting and put it directly into the ground in March. Willow bark was the souce of the first aspirin.

SEM pictures by Halbritter, H. In: PalDat - A palynological database.
https://www.paldat.org/pub/Salix_caprea/306378; accessed 2022-03-31 Polar and Equatorial

Verbenaceae - Verbena Family

Key words : Slightly irregular flowers with parts in fives and often wavy petals

The Verbenaceae are a family, commonly known as the verbena family or vervain family, of mainly tropical flowering plants. It contains trees, shrubs, and herbs notable for heads, spikes, or clusters of small flowers, many of which have an aromatic smell.

The Verbenaceae family includes some 35 genera and 1,200 species.

Economically important Verbenaceae include:

  - Lemon verbena (Aloysia triphylla), grown for aroma or flavoring
  - Verbenas or vervains (Verbena), some used in herbalism, others grown in gardens

The flowers are bisexual and slightly irregular with wavy petals. They are usually 5 united sepals and 5 united petals forming a tube with unequal lobes. Usually 4 stamens.
The ovary is superior and forms an equal number of chambers.
The fruit matures as 1 or 2 nutlets per carpel.
It also has squarish stems.

Species of importance to bees :

1. Common Name : Verbena
   Latin Name : *Verbena bonariensis*

Small tortoiseshell butterfly foraging on verbena in 2019

Verbena

Family :	Verbenaceae	Latin Name :	***Verbena bonariensis***
Common Name :	Verbena	Alternate Names :	Purpletop, South American vervain
Description :	Deciduous shrubs and trees	Flowering times :	July - November

Flowers : fragrant lavender to rose-purple flowers are in tight clusters located on terminal and axillary stems.

Leaves : are ovate to ovate-lanceolate with a toothed margin and grow up to 10 cm long.

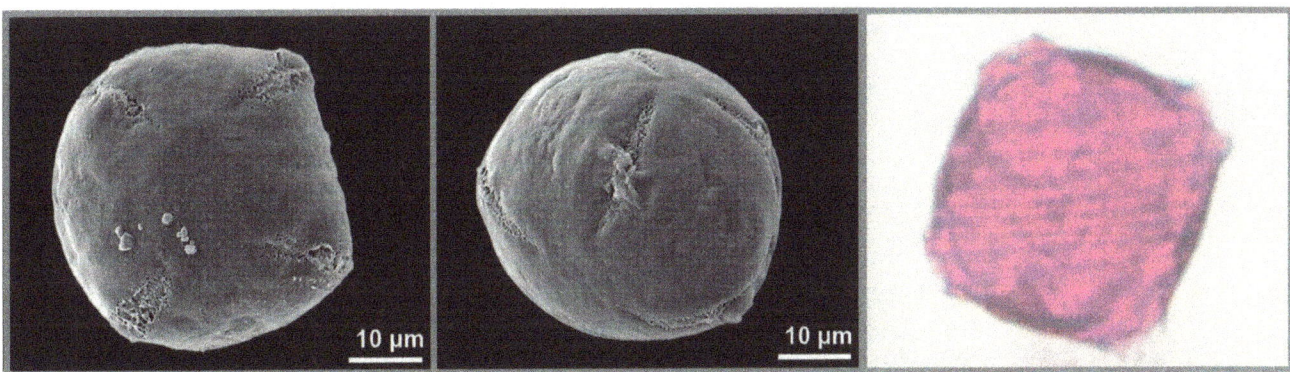

Pollen : Pollen loads are light brown. The pollen grains are 40 um in diameter. The pollen grains are round in shape. There are between 4 and 6 apertures which are furrows. The surface structure is smooth or indefinite and the exine appears to be of medium thickness. There appear to be granules or projections scattered on the apertures.

Nectar presentation : No honey crop.

Preferred habitat : Rough grasslands, scrub, roadsides, bare ground on well drained calcareous soils.

Particular special features : Attractive to bees and other nectar seeking insects such as butterflies. They are vigorously worked for nectar and pollen.

Wild verbena (verbena officinalis) has a history of medicinal use. It contains glycosides, tannins, bitters and volatile oils. Making it a sedative, diuretic, antispasmodic and bitter tonic. A tea is made from it to settle the stomache and relieve cold symptoms. However, the tea is bitter and too much of it can cause vomiting.

SEM pictures by Halbritter, H. In: PalDat - A palynological database.
https://www.paldat.org/pub/Verbena_bonariensis/302375; accessed 2022-03-31 Polar and Equatorial

Caprifoliaceae - Honeysuckle Family

Key words : Opposite leaves, with Aster-like blooms, irregular flowers and pithy stems

The Caprifoliaceae or honeysuckle family are dicotyledonous flowering plants consisting of about 860 species in 42 genera, with a nearly cosmopolitan distribution. Centres of diversity are found in eastern North America and eastern Asia, while they are absent in tropical and southern Africa.

The flowering plants in this family are mostly shrubs and vines : rarely herbs. They include some ornamental garden plants grown in temperate regions.

The leaves are mostly opposite with no stipules and may be either evergreen or deciduous.

The flowers are bisexual and slightly irregular, clustered in a dense head and often protected by spikey bracts.
There are 5 sepals and 4 or 5 untied petals plus 4 stamens.
The ovary is inferior but consistes of 2 united carpels whcih forms just 1 chamber producing a dry seed.

Species of importance to bees :

1. Common Name : Teasel
   Latin Name : **Dipsacus fullonum**

Tree bumblebee *Bombus hypnorum* foraging on teasel in 2019

Teasel

Family :	Caprifoliaceae	Latin Name :	***Dipsacus fullonum***
Common Name :	Teasel	Alternate Names :	Wild teasel or Fuller's teasel
Description :	Flowering plant and cultivar to 2 m	Flowering times :	July - August

Flowers : The inflorescence is a cylindrical array of lavender flowers which dries to a cone of spine-tipped hard bracts. It may be 10 cm long.

Leaves : lanceolate, 20-40 cm long and 3-6 cm broad, with a row of small spines on the underside of the midrib.

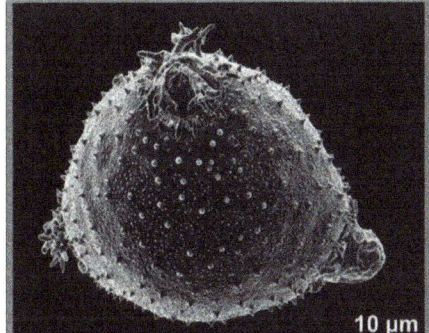
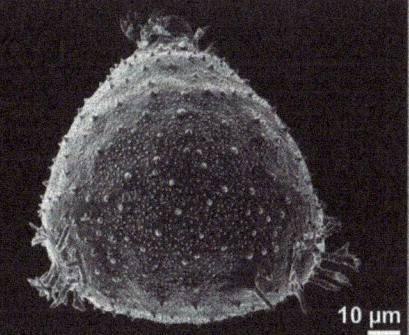
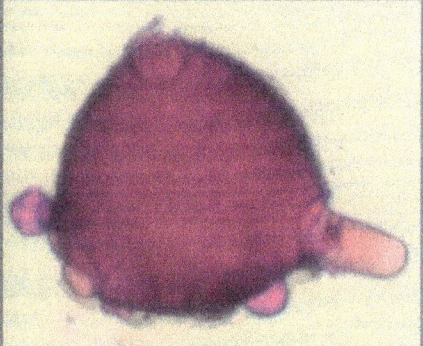

Pollen : Pollen loads are white. The pollen grains are 75 um in diameter. The pollen grains are round but triangular in cross-section. There are 3 apertures which are pores. The surface structure is made up of isolated dots due to spines or other projections. The exine is described as small or very small spines or warts. The aperture surface has some ornamentation.

Nectar presentation : No honey crop.

Preferred habitat : Rough grassland, roadsides, river banks and waste ground in rich soils that may be moist. However, they tolerate drought.

Particular special features : The flowering starts at the base of the flower head with only a few flowers open at a time. However, they freely offer nectar and are visited by a host of bee species. The flowering sequence moves steadily up the flowering head meaning that the flowers are available for several weeks.

The upper leaves of teasel have evolved to catch water and drown insects, absorbing the nutrients from the naturally decaying bodies. The spiny heads have been used in the wollen industry to produce a nap on wollen cloth (much loved by snooker players).

Also attractive to goldfinches - traditionally known as thistle-finches.

SEM pictures by Halbritter, H. In: PalDat - A palynological database.
https://www.paldat.org/pub/Dipsacus_fullonum/304567; accessed 2022-03-31 Polar and Equatorial

Lythraceae - Loosestrife Family

Key words : Twice the number of stamens as petals in 2 series (short and tall)

Lythraceae is a family of flowering plants, including 32 genera with about 620 species of herbs, shrubs and trees.

Lythraceae has a worldwide distribution, with most species in the tropics, but ranging into temperate climate regions as well.

The family is named after the type genus, Lythrum, the loosestrifes and also includes henna.

They have opposite or whorled leaves
They have regular bisexual flowers with 4, 6 or 8 sepals and the same number of petals.
There is typically twice as many stamens as petals forming 2 circles of different lengths.
It has a superior ovary forming an equal number of chambers forming a capsule with several to numerus seeds.

Species of importance to bees :

1. Common Name : Purple loosestrife
    Latin Name :   **Lythrum salicaria**

Flower spike of purple loosestrife

Key words : Twice the number of stamens as petals in 2 series (short and tall)

Purple loosestrife

Family :	Lythraceae	Latin Name :	***Lythrum salicaria***
Common Name :	Purple loosestrife	Alternate Names :	Black blood, Rebel weed.
Description :	Robust herbaceous perennial to 1m	Flowering times :	June - September

Flowers : star-shaped flowers in leafy racemes. Small vivid purplish-pink flowers 2cm wide in dense terminal spikes over a long period in summer.

Leaves : simple leaves in opposite pairs

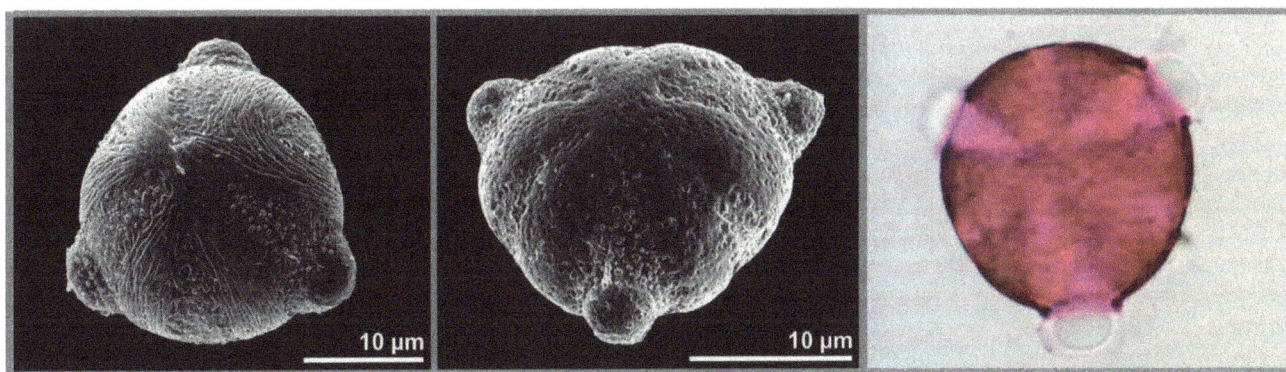

Pollen : Pollen loads are yellow - green. The pollen grains are 20 um in diameter. The pollen grains are round but triangular in cross-section. There are 3 apertures which are furrows with pores. The surface structure is smooth or indefinite and the exine appears to be medium in thickness. There does not appear to be any features on the surface of the apertures.

Nectar presentation : Not a significant crop. Light to dark colour with a strong aromatic sharp flavour.

Preferred habitat : Tall fens, reed swamp, marginal vegetation of lakes, rivers, ponds and canals.

Particular special features : It is an excellent bee plant and produces pollen and nectar in large quantities. It is visited by short and long-tongued bumblebees, solitary bees and honey bees.

SEM pictures by Halbritter, H. In: PalDat - A palynological database.
https://www.paldat.org/pub/Lythrum_salicaria/306279; accessed 2022-03-31 Polar x2

Onagraceae - Willowherb Family

Key words : Flower parts usually in fours including a four-lobed stigma

The Onagraceae are a family of flowering plants known as the willowherb family or evening primrose family.

They include about 650 species of herbs, shrubs, and trees in 17 genera. The family is widespread, occurring on every continent from boreal to tropical regions.

Some, particularly the willowherbs (Epilobium), are common weeds in gardens and rapidly colonise disturbed habitats in the wild.

The family is characterised by flowers with usually four sepals and petals; in some genera, such as Fuchsia, the sepals are as brightly coloured as the petals.

Flowers are regular bisexual with 4 separate sepals and 4 separate petals
There are an equal number or twice as many stamens as petals.
The ovary is inferior consisting of 4 united carpels.
The distinctive 4 parted stigma is the one thing to distinguish members of ths family.

Species of importance to bees :

1. Common Name : Rosebay willowherb
   Latin Name :   **Epilobium angustifolium**

White-tailed Bumblebee *Bombus lucorum* foraging on rosebay willowherb

Rosebay willowherb

Family :	Onagraceae	Latin Name :	*Epilobium angustifolium*
Common Name :	Rosebay willowherb	Alternate Names :	Fireweed, Chamerion or Chamaenerion
Description :	A perennial herbaceous plant	Flowering times :	July - August

Flowers : the flowers are 2 to 3 cm in diameter, slightly asymmetrical, with four magenta to pink petals and four narrower pink sepals behind. Note the 4 lobed stigma.

Leaves : are spirally arranged, entire, narrowly lanceolate, and pinnately veined.

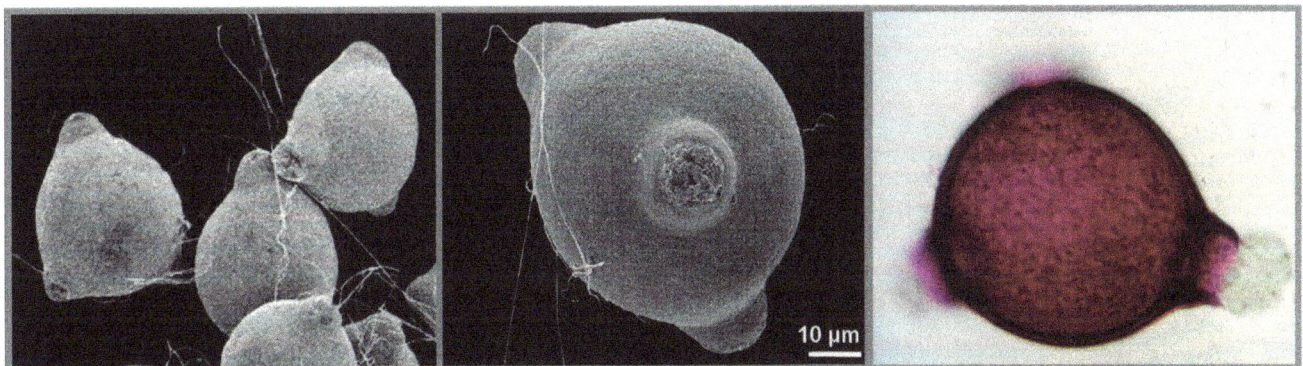

Pollen : Pollen loads are yellow - green. The pollen grains are 70 um in diameter. The pollen grains are round and somewhat triangular. There are 3 apertures which are pores. The surface structure is smooth or indefinite and the exine appe rs to be of medium thickness. There appears to be some ornamentation to the apertures, especially the edges. The fine fibres are called 'viscin threads' and are thought to be part of the adhesion method of the pollen grain to the stigma. I always think that the pollen grains remind me of inflated rubber gloves.

Nectar presentation : Medium honey crop. Very pale with fine granulation. Flavour mild and very sweet.

Preferred habitat : Forms dense stands on waste ground, roadsides, railways, heathland, woodland clearings, burnt ground (hence 'fireweed') and upland scree.

Particular special features : It is a good bee plant thanks to its long flowering season. It begins flowering at the bottom of the flower spike and this progresses upwards over a period of several weeks, producing pollen and nectar in abundance. It is attractive to honey bees, short and long-tongued bumblebees and solitary bees.

The male and female flowers appear on the same plant but a slightly different times. Near the top of a spike (young flowers) in the male phase and flowers near the base (older ones) in the female phase. Bees always go to the bottom of the spike first and work their way up. This means that they deliver their pollen load first and then move upwards until they collect pollen from the upper flowers. The bees are lured to the lower (female) flowers because they have more nectar than the upper ones.

The practice of burning the heather heathland can result in willoherb appearing which, in turn, can result in heather honey crops being 'spoilt' from the inclusion of willowherb honey producing a honey blend rather than pure heather honey.

It has a Pollen Coefficient of 0.3 which means it is 'under-represented' in honey.

SEM pictures by Halbritter, H. In: PalDat - A palynological database.
https://www.paldat.org/pub/Epilobium_angustifolium/304240; accessed 2022-03-31 Equatorial and multiple grains

Sapindaceae - Maple Family

Key words : Trees with large nut-like seeds encased in leathery translucent 'peel'

The Sapindaceae are a family of flowering plants in the order Sapindales known as the maple or soapberry family. It contains 138 genera and 1858 accepted species.

The Sapindaceae occur in temperate to tropical regions, many in laurel forest habitat, throughout the world.

Many are laticiferous, i.e. they contain latex, a milky sap, and many contain mildly toxic saponins with soap-like qualities in either the foliage and/or the seeds, or roots.

The flowers are bisexual, regular or slightly irregular grouped in a cluster at the end of a stem.
There are typically 5 untied sepals and 4 separate or basally attached petals plus 5 to 10 stamens.

The ovary is superior and consists of 3 united carpels. The ovary usually matures as a sinlge large nut-like seed. (Think horse chestnut)

Species of importance to bees :

1. Common Name : Horse chestnut
   Latin Name :   ***Aesculus hippocastanum***

2. Common Name : Maple
   Latin Name :   ***Acer* spp**

3. Common Name : Sycamore
   Latin Name :   ***Acer pseudoplatanus***

Tree Bumblebee *Bombus hypnorum* foraging on field maple in May 2020

Horse chestnut

Family :	Sapindaceae	Latin Name :	***Aesculus hippocastanum***
Common Name :	Horse chestnut	Alternate Names :	Conker tree
Description :	A large tree, growing to 40 m tall	Flowering times :	May - June

Flowers : are usually white with a yellow to pink blotch at the base of the petals. In erect panicles 10-30 cm tall with about 20-50 flowers on each panicle.

Leaves : are opposite and palmately compound, with 5-7 leaflets; each leaflet is 13-30 cm long, the whole leaf up to 60 cm across.

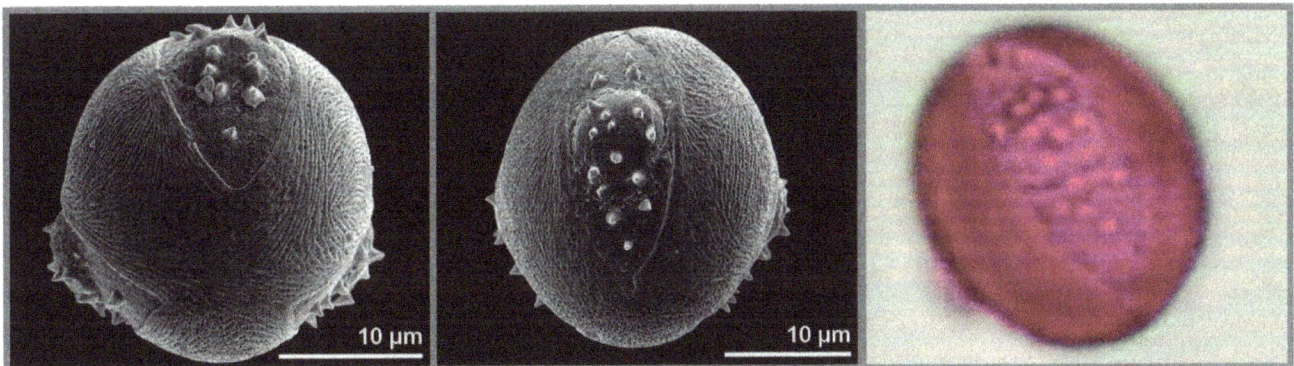

Pollen : Pollen loads are red - brown. The pollen grains are 20 um in diameter. The pollen grains are oval elongated in shape. There are 3 apertures which are furrows wth pores. The surface structure is smooth to striated and the exine appears to be thin. The surface of the apertures has scattered granules or projections.

Nectar presentation : Not a significant honey crop. Dark with rapid granulation

Preferred habitat : Parks, large gardens and village greens except in waterlogged conditions.

Particular special features : A common tree in the UK, it is well-worked by bees for nectar and pollen. The nectar guides on the petals start off yellow and then turn pink as the flowers are pollinated. This effectively turns the flower 'off' to the bees and enables more efficient pollination of the remaining flowers.

The sticky resinous buds are said to be a source of propolis collected by honey bees.

SEM pictures by Halbritter, H. In: PalDat - A palynological database.
https://www.paldat.org/pub/Aesculus_hippocastanum/304179; accessed 2022-03-31 Polar and Equatorial

Maple

Family :	Sapindaceae	Latin Name :	*Acer* **spp**
Common Name :	Maple	Alternate Name :	None
Description :	A genus of trees or shrubs	Flowering times :	April - May

Flowers : are regular, pentamerous (having parts arranged in groups of five), and borne in racemes, corymbs, or umbels with 4 or 5 petals about 1 - 6 mm long.

Leaves : are palmate veined and lobed, with 3 to 9 (rarely to 13) veins each leading to a lobe.

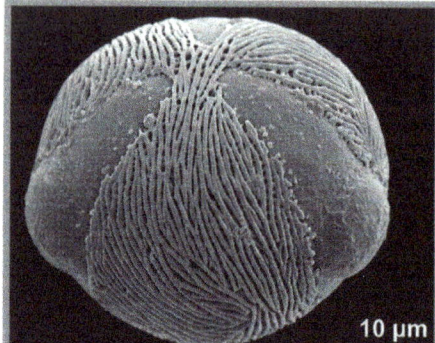

Pollen : Pollen loads are light to dark green. The pollen grains are 40 um in diameter. The pollen grains are round to triangular with 3 pores that are furrows. The surface structure is striated with the exine appearing to be medium in thckness. There are no apparent surface features on the apertures.

Nectar presentation : No honey crop.

Preferred habitat : Woodland and hedgerows on moist basic soils.

Particular special features : A good supply of pollen and nectar for early bees. Obviously being weather dependent but they can be worked by short and long-tongued bumblebees, solitary bees and honey bees.

In the spring of 2020, during the COVID-19 lockdown, I passed a maple on a daily walk and it was worked harder than any other tree, on that particular route, visited by a huge number of different insects.

SEM pictures by Halbritter, H. In: PalDat - A palynological database.
https://www.paldat.org/pub/Acer_palmatum/302550; accessed 2022-03-31 Polar and Oblique

Sycamore

Family : Sapindaceae
Common Name : Sycamore

Latin Name : *Acer pseudoplatanus*
Alternate Names : None

Description : A large deciduous, broad-leaved tree

Flowering times : April - June

Flowers : flowers are greenish-yellow and hang in dangling flowerheads called panicles. They produce copious amounts of pollen and nectar.

Leaves : leaves grow on long leafstalks and are large and palmate, with 5 large radiating lobes.

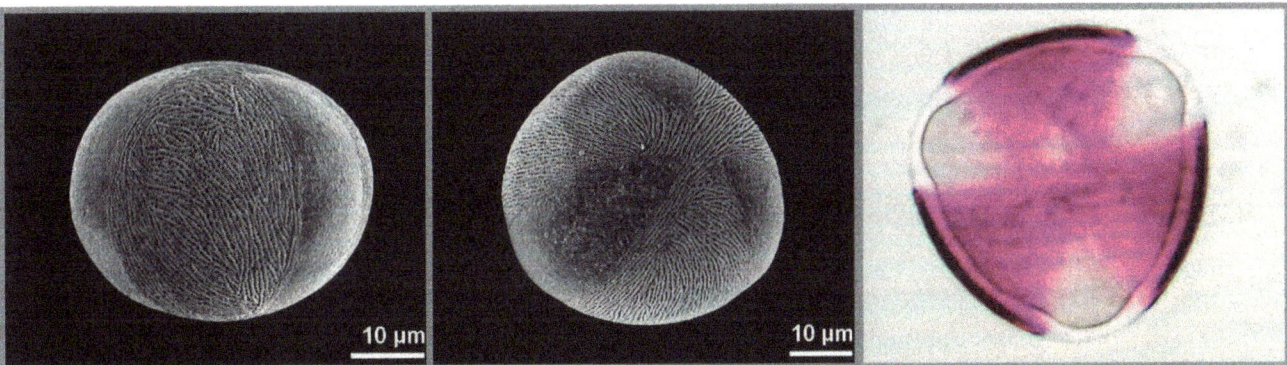

Pollen : Pollen loads are yellow - green. The pollen grains are 40 um in diameter. The pollen grains are oval to triangular in shape. There are 3 apertures which are furrows. The surface structure is striated and the exine is of meduim thickness. There are no apparent surface features on the apertures.

Nectar presentation : Medium honey crop. Light amber colour with fine granulation. It has a rank flavour that improves with age.

Preferred habitat : Parks, gardens, plantations, woods, roadsides and hedgerows.

Particular special features : The flowering period is relatively short but the flowers are a good surce of early nectar and pollen on which honey bee stocks can build-up. Sycamore is also visited by bumblebees and solitary bees.

SEM pictures by Sam, S. In: PalDat - A palynological database. https://www.paldat.org/pub/Acer_pseudoplatanus/304173; accessed 2022-03-31 Polar and Equatorial

Malvaceae - Mallow Family

Key words : Flowers with multiple stamens fused in a central column

Malvaceae, or the mallows, is a family of flowering plants estimated to contain 244 genera with 4225 known species.

Well-known members of economic importance include okra, cotton, cacao and durian.

There are also some genera containing familiar ornamentals, such as Alcea (hollyhock), Malva (mallow) and Lavatera (tree mallow).

Mallow leaves are usually alternate and palmately lobed.

The flowers are distinct and funnel shaped. They are regular and often surroundeed by several bracts. They have 3 to 5 partially united sepals and 5 separate petals.
There are numerous stamen that form a column around the pistil.

The ovary is superior and consists of usually 5 united carpels as indicated by the same number of styles.
The ovary matures as a capsule or rarely a winged seed or berry.

Many mallows contain natural gums called mucilage - muscilaginous.

Species of importance to bees :

1. Common Name : Lime
   Latin Name :   *Tilia* spp

2. Common Name : Common mallow
   Latin Name :   *Malva sylvestris*

3. Common Name : Musk mallow
   Latin Name :   *Malva moscata*

4. Common Name : Tree mallow
   Latin Name :   *Lavatera*

5. Common Name : Hollyhock
   Latin Name :   *Alcea rosea*

6. Common Name : Greek mallow
   Latin Name :   *Sidalcea* spp

Bumblebee on hollyhock

Lime

Family : Malvaceae
Common Name : Lime

Latin Name : ***Tilia* spp**
Alternate Names : Linden

Description : 30 species of trees or bushes

Flowering times : June - July

Flowers : flowers are white-yellow, five-petalled and hang in clusters of 2-5 and have a drooping habit.

Leaves : are dark green in colour, heart-shaped and flimsy and measure 6-10 cm in length.

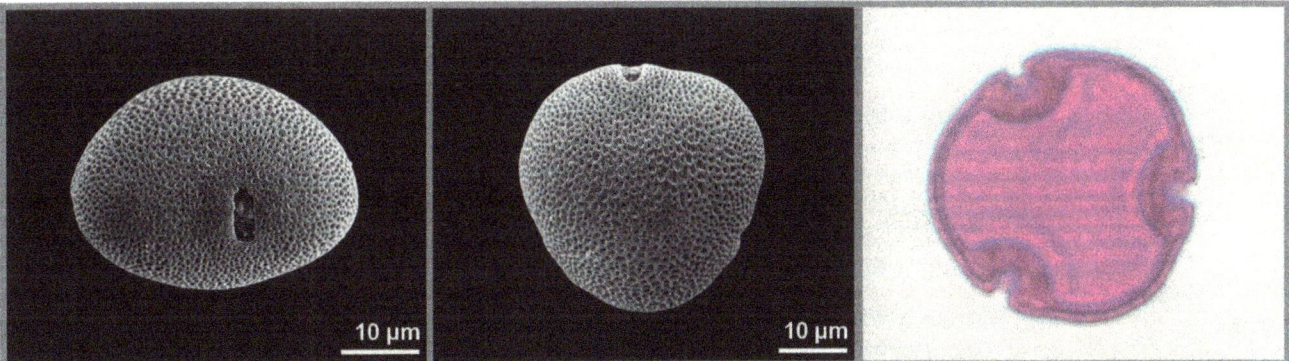

Pollen : Pollen loads are yellow. The pollen grains are 30 um in diameter. The pollen grains are oval to triangular in shape. There are 3 apertures which are furrows with pores. The surface structure is netted or pitted and the exine is of medium thickness. There are no apparent surface features on the apertures.

Nectar presentation : Can be a major honey crop. Very light colour with a fine slow granulation. It is said to taste of mint or minty'. Bees have been found dead or drunk under lime trees. Histroically, this was said to be caused by mannose (a sugar) that the bees cannot tolerate. However, more recent German research has disproved the mannose theory and the cause remains unexplained. It is also prone to honeydew which tastes like treacle. The bees, in this instance, are foraging on a secreted sugar solution from aphids which feed on some varieties of lime trees.

Preferred habitat : Parks, gardens, roadsides and wooded cliffs on a range of fertile soils except waterlogged.

Particular special features : The flowers of lime open at night and last about a week. They secrete nectar mainly during the morning. They produce large volumes of nectar and as such are of great value to urban beekeepers. They do not secrete nectar in cold conditions, the optimum temperature being around 20 degrees Celsius.
Nectar production in lime trees is said to be a little fickle and the reasons not well understood. The main theory relates to unfavourable temperatures i.e. being too low because of high winds. It is said that when the nights are cool and the days are warm and humid that the lime trees will produce a good nectar crop. This also affected by the age of the tree and also differs between the top and bottom of the tree. The lower branches producing more nectar as do younger trees.
it has a Pollen Coefficient of 10 which means it is 'under-represented' in honey.

SEM pictures by Halbritter, H. In: PalDat - A palynological database.
https://www.paldat.org/pub/Tilia_cordata/306424; accessed 2022-03-31 Polar and Equatorial

Common Mallow

Family :	Malvaceae	Latin Name :	*Malva sylvestris*
Common Name :	Common mallow	Alternate Name :	None
Description :	Showy flowers of bright mauve-purple	Flowering times :	June - September

Flowers : appear in axillary clusters of 2 to 4 and form irregularly and are elongated along the main stem. The flowers at the base open first. Petals are wrinkly to veined on the backs and are 15 to 25 mm long and 1 cm wide.

Leaves : are borne upon the stem, are roundish and have three to nine shallow lobes, each 2 to 4 cm long.

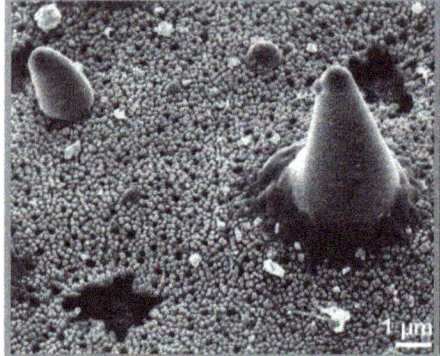 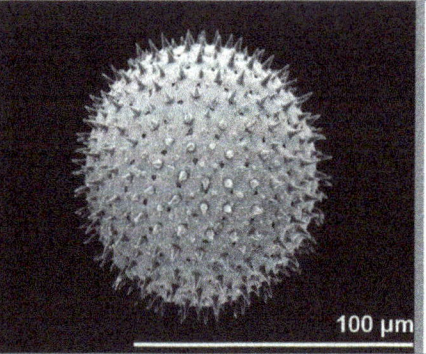 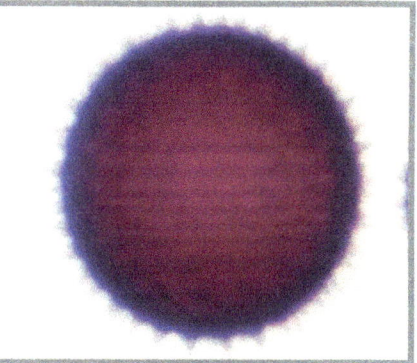

Pollen : Pollen loads are white - grey. The pollen grains are 100 um in diameter. The pollen grains are round. There are more than 12 apertures which are pores. The surface structure is granular with isolated dots due to spines or other projections. The exine is covered in long thin spines and there are no surface features on the apertures.

Nectar presentation : No honey crop.

Preferred habitat : Roadsides, banks, waste ground on well-drained soils.

Particular special features : The pollen grains are relatively massive being circa 100 microns in diameter compared to the 5 micron length of forget-me-not.

The flowers are very attractive to a wide variety of bee species from honey bees to bumblebees and solitary bees.

SEM pictures by Halbritter, H. In: PalDat - A palynological database.
https://www.paldat.org/pub/Malva_sylvestris/304838; accessed 2022-03-31 Polar and Exine surface

Musk Mallow

Family :	Malvaceae	Latin Name :	*Malva moscata*
Common Name :	Musk mallow	Alternate Names :	None
Description :	Herbaceous perennial plant to 1m	Flowering times :	June - September

Flowers : are produced in clusters in the leaf axils, each flower 3-5 cm diameter, with five bright pink petals. They have a distinctive musky odour.

Leaves : leaves are alternate, 2-8 cm long and 2-8 cm broad, palmately lobed with 5 to 7 lobes.

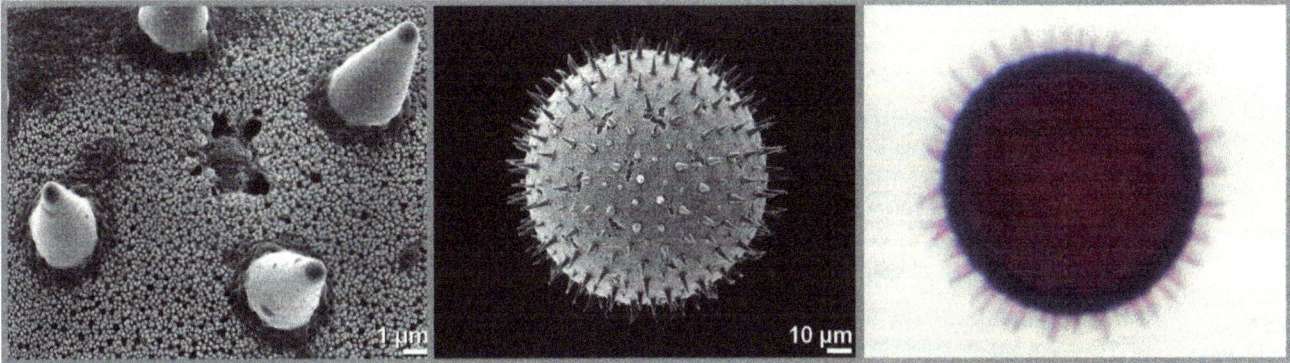

Pollen : Pollen loads are white. The pollen grains are 90 um in diameter. The pollen grains are round. There are more than 12 apertures which are pores. The surface structure is granular with isolated dots due to spines or other projections. The exine is covered in long thin spines and there are no surface features on the apertures.

Nectar presentation : No honey crop.

Preferred habitat : Roadsides, hedge banks, pastures and field borders on well-drained soils.

Particular special features : As for common mallow

SEM pictures by Halbritter, H. In: PalDat - A palynological database.
https://www.paldat.org/pub/Malva_moschata/304197; accessed 2022-03-31 Polar and Aperture

Tree Mallow

Family : Malvaceae
Common Name : Tree mallow

Latin Name : *Lavatera*
Alternate Name : Rose mallows, Royal mallows

Description : Genus 25 species of flowering plants

Flowering times : June - September

Flowers : are conspicuous, 4-12 cm diameter, with five white, pink or red petals; they are produced in terminal clusters.

Leaves : the leaves are spirally arranged, and palmately lobed.

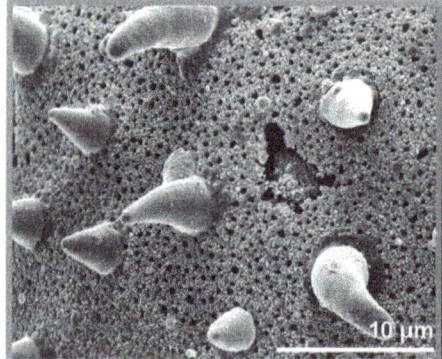
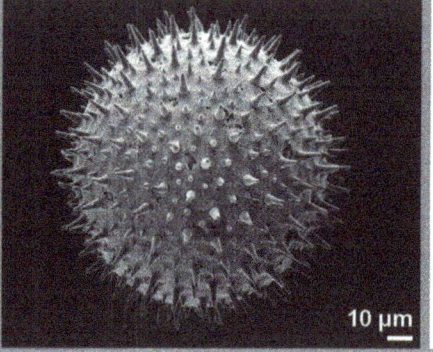
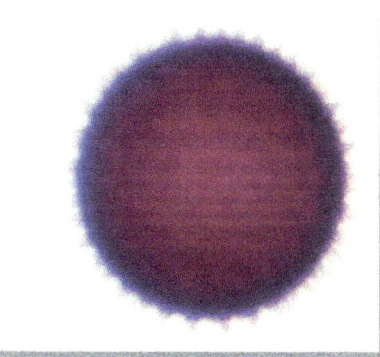

Pollen : Pollen loads are white. The pollen grains are 100 um in diameter. The pollen grains are round. There are more than 12 apertures which are pores. The surface structure is granular with isolated dots due to spines or other projections. The exine is covered in long thin spines and there are no surface features on the apertures.

Nectar presentation : No honey crop.

Preferred habitat : Cliffs, rocks, waste ground close to the sea.

Particular special features : Yeilds nectar and pollen and are good for honey bees bumblebees and solitary bees.

SEM pictures by Halbritter, H. In: PalDat - A palynological database.
https://www.paldat.org/pub/Lavatera_thuringiaca/304202; accessed 2022-03-31 Polar and Aperture

Hollyhock

Family : Malvaceae
Common Name : Hollyhock

Latin Name : *Alcea rosea*
Alternate Names : None

Description : Annual, biennial, or perennial plants

Flowering times : June - August

Flowers : may be solitary or arranged in fascicles (crowded together) or racemes. The notched petals are usually over 3 cm wide and may be pink, white, purple, or yellow.

Leaves : the leaf blades are often lobed or toothed, and are borne on long petioles.

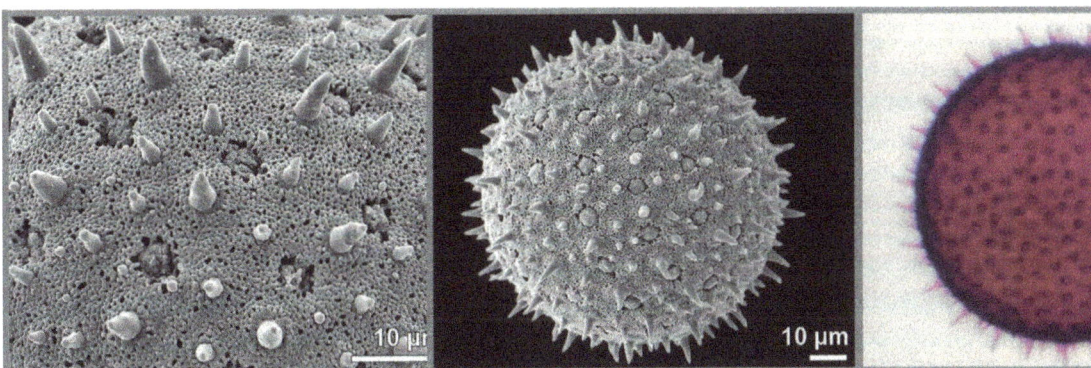

Pollen : Pollen loads are brown. The pollen grains are 140 um in diameter. The pollen grains are round. There are more than 12 apertures which are pores. The surface structure is granular with isolated dots due to spines or other projections. The exine is covered in long thin spines and there are no surface features on the apertures.

Nectar presentation : No honey crop.

Preferred habitat : Cracks in pavements in a sunny position with rich soils.

Particular special features : The hollyhock is good source of late-season pollen. They are also worked somewhat for nectar by honey bees and bumblebees. There large tap root makes them drought tollerant.

SEM pictures by Halbritter, H. In: PalDat - A palynological database.
https://www.paldat.org/pub/Alcea_ficifolia/300582; accessed 2022-03-31 Polar and Pollen surface

Greek Mallow

Family :	Malvaceae	Latin Name :	*Sidalcea* **spp**
Common Name :	Greek mallow	Alternate Name :	Checkerblooms, Checkermallows, or Prairie mallows.
Description :	Annual or perennial flowering plants	Flowering times :	July - September

Flowers : produce erect flowering stems, with 5-petalled flowers in terminal racemes, in shades of pink, white and purple

Leaves : clumps of toothed basal leaves

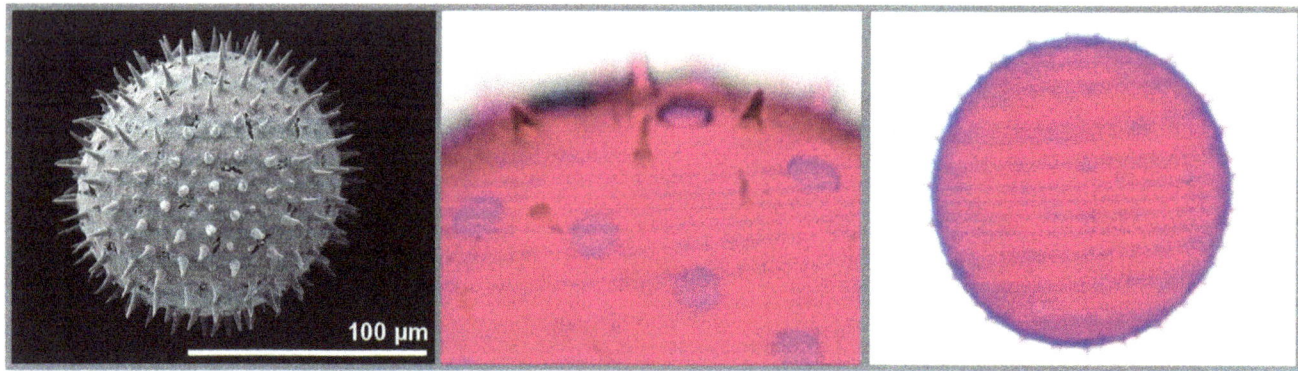

Pollen : Pollen loads are light brown. The pollen grains are 100 um in diameter. The pollen grains are round. There are more than 12 apertures which are pores. The surface structure is granular with isolated dots due to spines or other projections. The exine is covered in long thin spines and there are no surface features on the apertures.

Nectar presentation : No honey crop.

Preferred habitat : Flower borders

Particular special features : Most useful for pollen but nectar is collected too. Visited by Bumblebees, solitary bees and honey bees.

SEM pictures by Halbritter, H. In: PalDat - A palynological database.
https://www.paldat.org/pub/Malva_alcea/304198; accessed 2022-03-31 Polar

Brassicaceae - Mustard or Cabbage Family

Key words : 4 petals and 6 stamens (4 tall and 2 short)

Brassicaceae or Cruciferae is a medium-sized and economically important family of flowering plants commonly known as the mustards, the crucifers, or the cabbage family.

Most are herbaceous plants but there are some shrubs.

The mustards are identified by the flowers. There are 4 sepals, 4 petals and 6 stamens (4 long and 2 short). The petals are often arranged like the letters X ot H. They may also be deeply split appearing as if there are 8 petals.

The ovary is superior consisting of 2 untied carpels. It matures as a pod.

The crushed leaves usually smell like mustard.

Species of importance to bees :

1. Common Name : Oilseed Rape
   Latin Name : **Brassica napus**

2. Common Name : Charlock
   Latin Name : **Sinapis arvensis**

3. Common Name : Aubretia
   Latin Name : **Aubretia deltoidia**

4. Common Name : Sweet Alison
   Latin Name : **Lobularia maritima**

Photos above and below : Hopkinson family inspecting a rape field pre 1972 when it was grown for lubricating oil prior to the new strains being introduced.

Geoff Hopkinson

Oilseed Rape

Family : Brassicaceae
Common Name : Oilseed Rape
Description : A yellow flowering field crop

Latin Name : **Brassica napus**
Alternate Name : Rapeseed, Rape, OSR, Swede Rape
Flowering times : April - May

Flowers : yellow, approx. 1-1.5 cm across; petals four, 10-15 mm long.

Leaves : alternate, lowest stalked, upper stalkless. Blade glabrous, bluish green, basal leaves shallow-lobed, upper leaves entire and lanceolate.

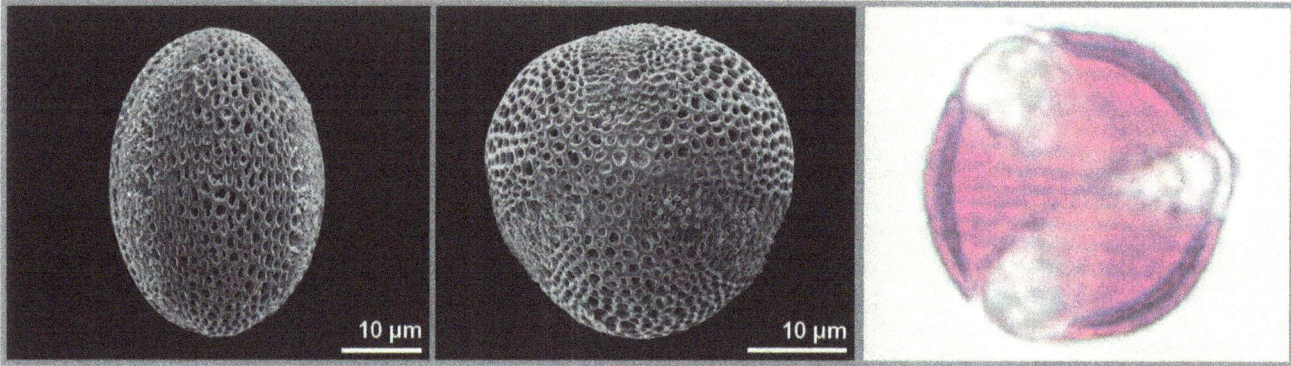

Pollen : Pollen loads are light yellow-green. The pollen grains are 30 um in diameter. The pollen grains are oval but triangular in cross-section. There are 3 apertures which are furrows. The surface structure is netted or pitted and the exine is of medium thickness with spaced rods. There are no apparent surface features on the apertures.

Nectar presentation : Major honey crop. Light colour with very rapid and fine granulation. Flavour is mild and sweet. The high glucose content is responsible for the rapid granulation. This can catch beekeepers out and the honey can set hard in the comb. Remove the supers before the honey is capped but check the water content using a sharp shake over the hive - if nectar comes out then it is not ready to take. If no nectar comes out then it is about 'ripe' and can be extracted. Remember to check with a refractometer to be sure.

Preferred habitat : Arable weed of roadsides and disturbed ground.

Particular special features : Winter sown rape is sown in August or September and flowers the following April to May and the seed crop is harvested in July or August. Spring sown rape flowers from June to July.

There are dark green nectaries at the base of the flower which are hard to reach for honey bees. They access the nectaries between the bases of the individual petals in a process known as 'base working'. For the flowers to yield nectar the air temperature needs to be above 16 degrees Celsius. This can be a problem because autumn sown rape will be in flower in April and the spring temperatures may not be high enough, not only for the plant to secrete nectar but for the bees to be able to fly to obtain the nectar. To make matters worse, the plant secretes most of its nectar in the morning betwwen 7 am and 10 am when the air temperature has not really had time to rise too much.

It has a Pollen Coefficient of 150 which means it is 'over-represented' in honey.

SEM pictures by Diethart, B. In: PalDat - A palynological database.
https://www.paldat.org/pub/Brassica_napus/303973; accessed 2022-03-31 Polar and Equatorial

Charlock

Family :	Brassicaceae	Latin Name :	*Sinapis arvensis*
Common Name :	Charlock	Alternate Names :	Field mustard or Wild mustard
Description :	An annual or winter annual plant	Flowering times :	May - August

Flowers : the inflorescence is a raceme made up of yellow flowers having four petals.

Leaves : are petiolate (stalked) with a length of 1-4 cm.

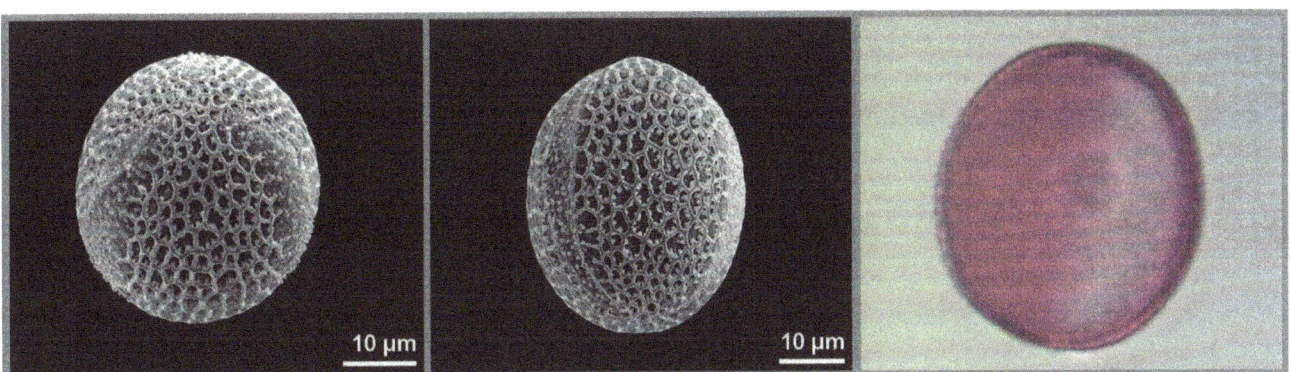

Pollen : Pollen loads are light yellow-green. The pollen grains are 30 um in diameter. The pollen grains are oval and have 3 apertures wich are furrows. The surface structure is netted or pitted and the exine is of medium thickness with spaced rods. There are no apparent surface features on the apertures.

Nectar presentation : Not a significant crop. Light to amber colour. Rapid fine granulation. Mild in flavour with a slight sharpness (like mustard). It is an excellent 'seed' honey for crops such as white clover.

Preferred habitat : Roadsides, waste ground, usually on calcareous soils.

Particular special features : The seeds can lie dormant in soil for up to 30 years and large areas of charlock can appear following deep ploughing. The flowers supply both nectar and pollen to both short and long-tongued bees including honey bees, bumblebbes and solitary bees.

SEM pictures by Bombosi, P. In: PalDat - A palynological database.
https://www.paldat.org/pub/Sinapis_alba/306408; accessed 2022-03-31 Polar and Equatorial

Aubretia

Family :	Brassicaceae	Latin Name :	*Aubretia deltoidia*
Common Name :	Aubretia	Alternate Name :	Lilacbush, Purple rock cress and Rainbow rock cress
Description :	A small herbaceous perennial	Flowering times :	March - May

Flowers : the showy inflorescence bears small flowers with four lavender to deep pink petals.

Leaves : green spoon-shaped to oval-shaped leaves, some of which are lobed.

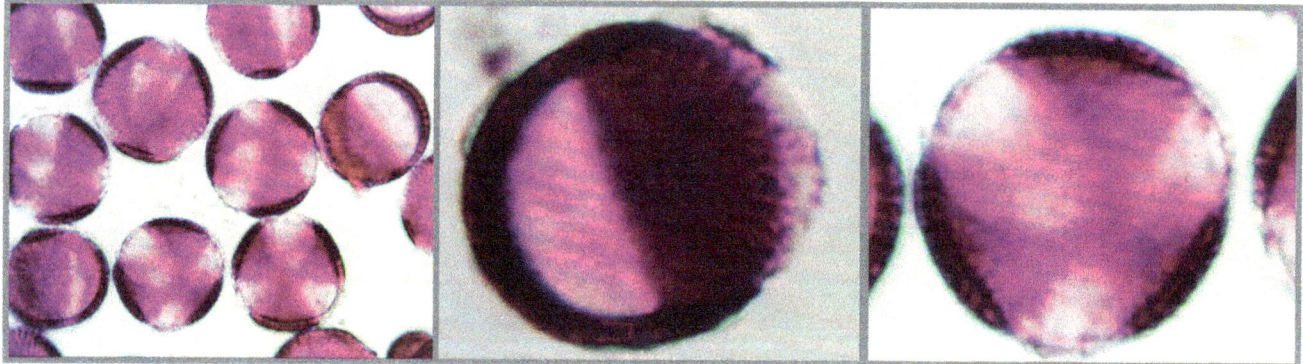

Pollen : Pollen loads are light green. The pollen grains are 25 um in diameter. The pollen grains are round in shape and have 3 apertures which are furrows. The surface structure is netted or pitted and the exine is of medium thickness with spaced rods. There are no apparent surface features on the apertures.

Nectar presentation : No honey crop.

Preferred habitat : Scree on basic soils or chalk and limestone rocks.

Particular special features : A very good source of early pollen and nectar to lots of bee species. This includes honey bees, short and long-tongued bumblebees and solitary bees.

Sweet Alison

Family :	Brassicaceae	Latin Name :	*Lobularia maritima*
Common Name :	Sweet Alison	Alternate Names :	Sweet alyssum or just alyssum.
Description :	Annual plant / short-lived perennial	Flowering times :	July - September

Flowers : are about 5 mm in diameter, sweet-smelling, with an aroma similar to that of honey, with four white rounded petals (or pink, rose-red, violet or lilac) and four sepals.

Leaves : are 1-4 mm long and 3-5 mm, broad, alternate, sessile, quite hairy, oval to lanceolate, with an entire margin.

Pollen : Pollen loads are yellow. The pollen grains are 15 um in diameter. The pollen grains are oval in shape with 3 apertures which are furrows. The surface structure is netted or pitted and the exine is of medium thickness with spaced pods. There are no apparent surface features on the apertures.

Nectar presentation : No honey crop.

Preferred habitat : Arable fields, tracks, waste ground on sandy soils.

Particular special features : Apparenty edible, although the flowers give off a distinctly honey aroma they are not profuse nectar producers. Bees of various species visit the flowers but mainly for pollen. These include, honey bees, bumblebees and solitary bees.

SEM pictures by Halbritter, H. In: PalDat - A palynological database.
https://www.paldat.org/pub/Lobularia_maritima/306118; accessed 2022-03-31 Polar and Equatorial

Rutaceae - Citrus Family

Key words : Trees or shrubs, frequently aromatic, with 4 or 5 sepals and petals

The Rutaceae are a family, commonly known as the rue or citrus family, of flowering plants, usually placed in the order Sapindales.

Species of the family generally have flowers that divide into four or five parts, usually with strong scents. They range in form and size from herbs to shrubs and large trees.

The most economically important genus in the family is Citrus, which includes the orange, lemon, grapefruit, and lime.

Boronia is a large Australian genus, some members of which are plants with highly fragrant flowers and are used in commercial oil production. About 160 genera are in the family Rutaceae.

Species of importance to bees :

1. Common Name : Chinese Bee Tree
   Latin Name : *Tetradium daniellii*

2. Common Name : Mexican Orange Blossom
   Latin Name : *Choisya ternata*

Honey bee foraging on choisya

Chinese Bee Tree

Family :	Rutaceae	Latin Name :	*Tetradium daniellii*
Common Name :	Chinese Bee Tree	Alternate Names :	Euodia, Korean evodia, Bee-bee tree
Description :	A large deciduous tree to 30 m	Flowering times :	July - August

Flowers : are produced in domed, terminal corymbs to 15 cm across. Flower colour - White - with yellow anthers.

Leaves : pinnate leaves to 40 cm or more long, each with up to 11 elliptic, oval or lance-shaped, glossy, dark-green leaflets, turning yellow in autumn.

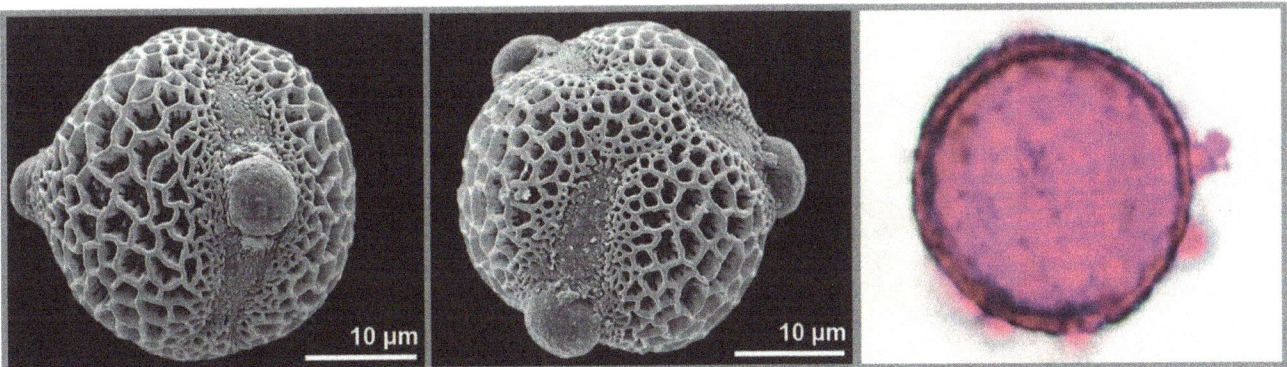

Pollen : Pollen loads are yellow. The pollen grains are 30 um in diameter. The pollen grains are essentially round in shape with 3 aperutes which are furrows with pores. The surface structure is netted or pitted and the exine is of medium thickness. There are no apparent surface features on the apertures.

Nectar presentation : No honey crop.

Preferred habitat : From Asia, specifically China and Korea, it prefers loamy soils and is not cold-hardy.

Particular special features : To experience this tree in the height of summer is amazing. There is a mature tree circa 30 metres tall in St. Andrews Botanic Garden. When I visited, the canopy was buzzing with insects including bees, flies wasps and hornets.

An oil from the fruit is edible and used in cooking and also for hair products.

An interesting tree that has been championed, and grown from seed, for many years by Geoff Hopkinson BEM NDB.

SEM pictures by Halbritter, H. In: PalDat - A palynological database.
https://www.paldat.org/pub/Tetradium_daniellii/306188; accessed 2022-03-31 Polar and Equatorial

Mexican Orange Blossom

Family :	Rutaceae	Latin Name :	*Choisya ternata*
Common Name :	Mexican Orange Blossom	Alternate Name :	Mexican orange
Description :	Evergreen shrub, growing up to 3 m	Flowering times :	May - June

Flowers : the white flowers are scented, appearing in spring (sometimes with limited repeat flowering in autumn).

Leaves : have three leaflets and are aromatic.

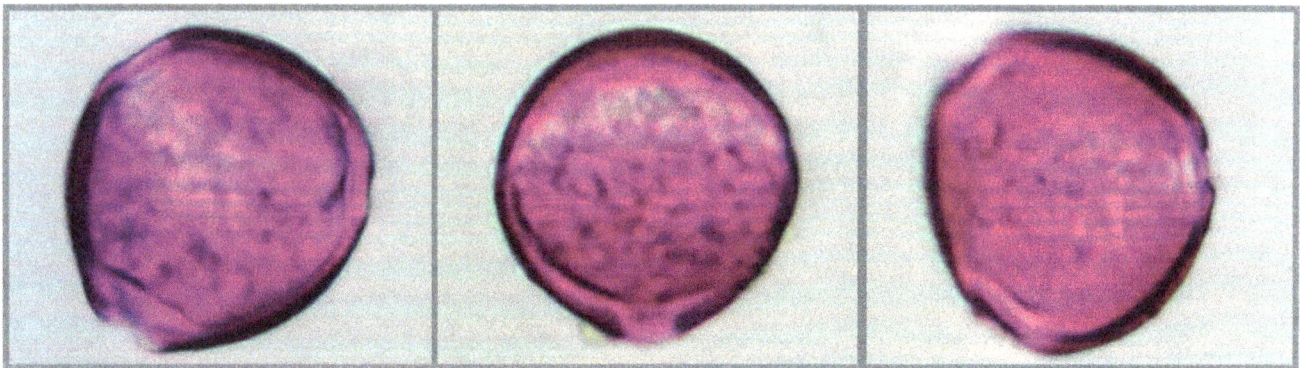

Pollen : Pollen loads are yellow. The pollen grains are 30 um in diameter. The pollen grains are essentially round in shape and there are 3 apertures which are pores. The surface structure is smooth or indefinite and the exine appears to be of medium thickness. There are no apparent features on the surface of the apertures.

Nectar presentation : No honey crop.

Preferred habitat : Originating from Mexico it is drought tollerant and prefers well-drained soils.

Particular special features : Honey bees love this plant. In the photograph opposite, you can see the bee hives in the background. When in flower, this bush is covered in honey bees from these hives.

Geoff Hopkinson BEM NDB in his garden enjoying his Choisya in full bloom. Note the bee hives in the background. The choisya was covered with honey bees.

Plumbaginaceae - Leadwort Family

Key words : Perfect flowers in salt-rich marshes and sea coasts

Plumbaginaceae is a family of flowering plants, with a cosmopolitan distribution. The family is sometimes referred to as the leadwort family or the plumbago family.

Most species in this family are perennial herbaceous plants, but a few grow as shrubs.

The plants have perfect flowers and are pollinated by insects. They are found in many different climatic regions, from arctic to tropical conditions, but are particularly associated with salt-rich marshes, and sea coasts.

Species of importance to bees :

1. Common Name : Sea lavender
   Latin Name :   *Limonium vulgare*

Sea lavender on salt marshes in Ichenor, West Sussex

Sea Lavender

Family : Plumbaginaceae
Common Name : Sea lavender

Latin Name : *Limonium vulgare*
Alternate Names : Common sea-lavender, Sea thrift

Description : Herbaceous perennial plants

Flowering times : July - August

Flowers : are produced on a branched panicle or corymb, the individual flowers are small (4-10 mm long). The flower colour is pink or violet to purple in most species.

Leaves : The leaves are simple, entire to lobed, and from 1-30 cm long and 0.5-10 cm broad.

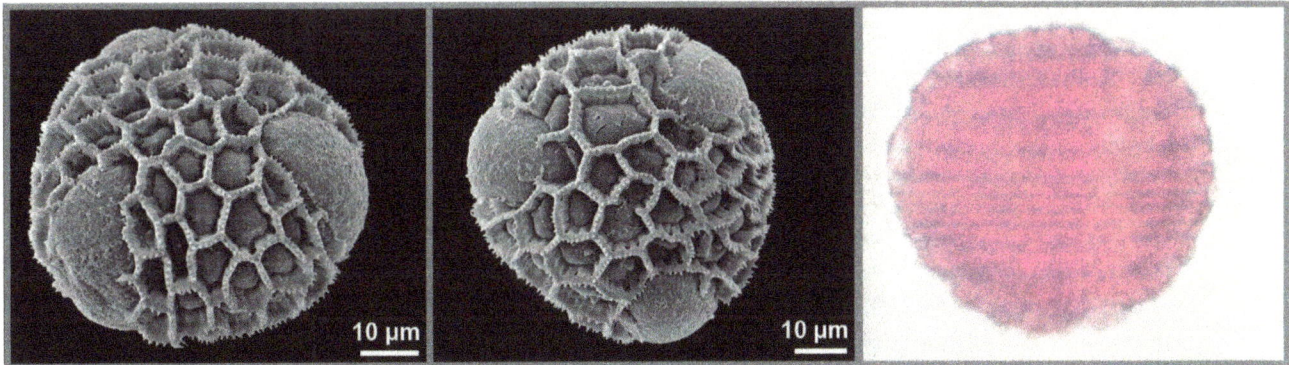

Pollen : Pollen loads are light yellow. The pollen grains are 50 um in diameter. The pollen grains are oval but triangular in cross-section. There are 3 apertures which are pores. The surface structure is netted or pitted and the exine appears to be thin. There are no apparent surface features on the apertures.

Nectar presentation : No honey crop.

Preferred habitat : Intermediate zone of inter-tidal salt marsh.

Particular special features : If it occurs in abundance then it can be a good late source of nectar for honey bees, bumblebees and solitary bees.

SEM pictures by Halbritter, H. In: PalDat - A palynological database.
https://www.paldat.org/pub/Limonium_cancellatum/301181; accessed 2022-03-31 Polar and Equatorial

Balsaminaceae - Balsam Family

Key words : Delicate, jiucy plants with irregular flowers

The Balsaminaceae (commonly known as the balsam family) are a family of dicotyledonous plants, comprising two genera:
Impatiens, which consists of 1000+ species, and
Hydrocera, consisting of 1 species.

The flowering plants may be annual or perennial. They are found throughout temperate and tropical regions, primarily in Asia and Africa, but also North America and Europe.

Notable members of the family include jewel weed and busy Lizzie.

In the flowers there are 3 (sometimes 5) petal-like sepals of unequal size. The lowest one forms a nectar filled spur. There are 5 petals, 2 are fused. There are 5 stamens.
The superior ovary has 5 fused carpels forming an equal number of chambers. Each carpel prduces 2 to numerous seeds.
In many species the ovary matures as a capsule which explodes when touched - hence the family is sometimes called "Touch-me-not"

Species of importance to bees :

1. Common Name : Himalayan balsam
   Latin Name :   ***Impatiens glandulifera***

Tree Bumblebee *Bombus hypnorum* foraging on himalayan balsam

Himalayan balsam

Family : Balsaminaceae
Common Name : Himalayan balsam

Description : Annual plant native to the Himalayas

Latin Name : **_Impatiens glandulifera_**
Alternate Names : Policeman's Helmet, Bobby Tops, Copper Tops, Gnome's Hatstand, Jewel weed.
Flowering times : June - October

Flowers : are pink (sometimes white), with a hooded shape, 3 to 4 cm tall and 2 cm broad; the flower shape has been compared to a policeman's helmet.

Leaves : lanceolate leaves 5 to 23 cm long. The crushed foliage has a strong musty smell.

Pollen : Pollen loads are white. The pollen grains are 20 x 10 um. The pollen grains are oval oblongs with 4 apertures which are furrows. The surface structure is smooth or indefinite and the exine is thin. There are no apparent surface features on the apertures.

Nectar presentation : Not a significant crop. Light in colour, mild in flavour and very watery.

Preferred habitat : Riversides, ditches, wet woodlands, very invasive forming dense stands.

Particular special features : This is an excellent bee plant and is visited by honey bees, short and long-tongued bumblebees and solitary bees. A profuse nectar producer that is obvious to the beekeeper when the bees are visiting it. The anthers are in the top of the flower and dust the pollen onto the backs of the bees as they access the nectaries in the base of the flower. The white spots are an excellent indication of the bees foraging on balsam plants. However, they can cause some worry, as I have had phone calls and been sent photographs of bees with this 'affliction' from new beekeepers concerned with what their bees were suffering from.

The seeds are apparently edible and it can be used as a poultice for skin irritations such as athlete's foot and bee stings.

Interestingly, they are originally from the Himalayas and as such are classed as non-native invasive species. This prohibits them from being planted or distributed to the wild.

SEM pictures by Halbritter, H. In: PalDat - A palynological database.
https://www.paldat.org/pub/Impatiens_glandulifera/306259; accessed 2022-03-31 Polar and Multiple grains

Ericaceae

Key words : Often evergreen, with bell-shaped flowers and flower parts in fives

The Ericaceae are a family of flowering plants, commonly known as the heath or heather family, found most commonly in acid and in fertile growing conditions.

The family is large, with c. 4250 known species spread across 124 genera, making it the 14th most species-rich family of flowering plants.

The many well-known and economically important members of the Ericaceae include the cranberry, blueberry, huckleberry, rhododendron (including azaleas), and various common heaths and heathers.

Species of importance to bees :

1. Common Name : Ling Heather
   Latin Name : **Calluna vulgaris**

2. Common Name : Bell Heather
   Latin Name : **Erica cinerea**

3. Common Name : Cross-leaved Heath
   Latin Name : **Erica tetralix**

4. Common Name : Winter Heath
   Latin Name : **Erica carnea**

5. Common Name : Darley Dale Heath
   Latin Name : **Erica darleyensis**

6. Common Name : Rhododendron
   Latin Name : **Rhododendron ponticum**

7. Common Name : Calico bush
   Latin Name : **Kalmia latifolia**

8. Common Name : Blueberry
   Latin Name : **Vaccinium corymbosum**

9. Common Name : Strawberry tree
   Latin Name : **Arbutus unedo**

10. Common Name : Pieris
    Latin Name : **Pieris spp**

11. Common Name : Bilberry
    Latin Name : **Vaccinium myrtillus**

12. Common Name : Cranberry
    Latin Name : **Vaccinium oxycoccos**

Buff-tailed Bumblebee *Bombus terrestris* foraging on Rhododendron ponticum

Ling Heather

Family :	Ericaceae	Latin Name :	*Calluna vulgaris*
Common Name :	Ling Heather	Alternate Name :	Common heather, Ling, or Heather
Description :	Low-growing perennial shrub	Flowering times :	July - Spetember

Flowers : normally mauve, but white-flowered plants also occur occasionally. They are terminal in racemes with sepal-like bracts at the base.

Leaves : small scale-leaves (less than 2-3 mm long) borne in opposite and decussate (cross or intersect each other to form an X) pairs.

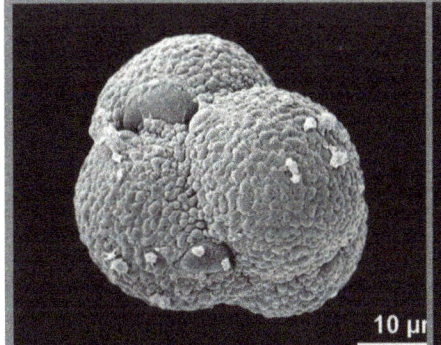
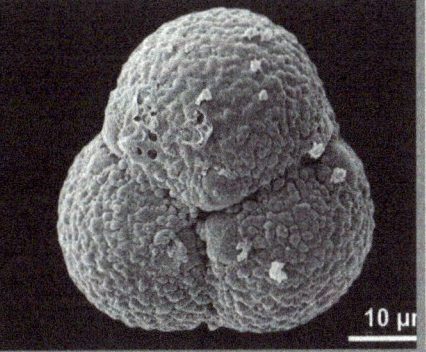
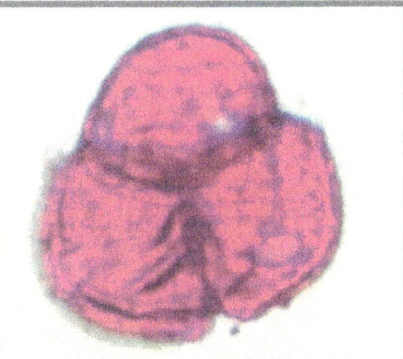

Pollen : Pollen loads are grey-brown. The tetrads are 40 um in diameter. The pollen grains are grouped together in groups of 4 in a tetrahedral configuration called 'tetrads'. The pollen grains are irregularly round with 3 aperutres which are furrows with pores. The surface structure is smooth or indefinite and the exine is of meduim thickness. There appears to be thickened or projecting edges to apertures.

Nectar presentation : Major honey source. Reddish amber colour. Slow coarse granulation. Slightly bitter flavour with a strong aroma from floral to medicinal. It makes good comb honey with particularly white cappings. The honey is thixatropic which means it is in a gel-like state until it is agitated then it beomes liquid. This means that it cannot be spun out of the comb in the usual way and needs to be agitiated by special needles first or else it is pressed from the comb. Pressing adds air bubbles that are retained in the honey.

Preferred habitat : Heaths, moors, bogs and open birch, pine and oak woodland on acid soils.

Particular special features : It is said that the best honey yields are from young plants and that heather moors need to periodically 'burnt' off to facilitate the new young plants replacing the older ones. Young shoots are better nectar producers than woody stems. However, the soil moisture and air temperature will also have a bearing on nectar yields. The pollen is also collected. Ling heather is visited by honey bees, bumblebees and solitary bees.

It has a Pollen Coefficient of 12 which means it is 'under-represented' in honey.

SEM pictures by Halbritter, H. In: PalDat - A palynological database.
https://www.paldat.org/pub/Calluna_vulgaris/304299; accessed 2022-03-31 Polar and Equatorial

Bell Heather

Family :	Ericaceae	Latin Name :	*Erica cinerea*
Common Name :	Bell Heather	Alternate Names :	None
Description :	Low, spreading shrub	Flowering times :	July - September

Flowers : are bell-shaped, purple (rarely white), 4-7 mm long.

Leaves : fine needle-like leaves 4-8 mm long arranged in whorls of three.

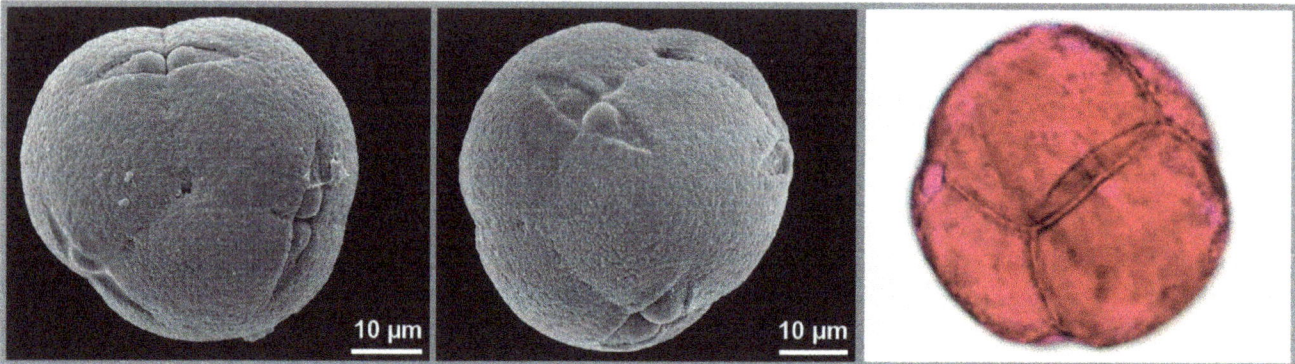

Pollen : Pollen loads are white-grey. The tetrads are 40 um in diameter. The pollen grains are grouped together in groups of 4 in a tetrahedral configuration called 'tetrads'. The pollen grains are irregularly round with 3 aperutres which are furrows with pores. The surface structure is smooth or indefinite and the exine is of meduim thickness. There appears to be thickened or projecting edges to apertures.

Nectar presentation : Major honey source. Port wine colour. Rapid granulation. Distinctive pronounced flavour.

Preferred habitat : Dry heathland, open woodland, maritime heaths on acid well-drained soils.

Particular special features : The bell-shaped corolla is about 5mm in length which allows access to honey bees and short-tongued bumblebees. However, the bumblebees still tend to pierce the corolla at the base through which they steal the nectar. This then enables honey bees to do the same.

It has a Pollen Coefficient of 10 which means it is 'under-represented' in honey.

SEM pictures by Halbritter, H. In: PalDat - A palynological database.
https://www.paldat.org/pub/Erica_cerinthoides/303004; accessed 2022-03-31 Polar and Equatorial

Cross-leaved Heath

Family :	Ericaceae	Latin Name :	*Erica tetralix*
Common Name :	Cross-leaved Heath	Alternate Name :	None
Description :	Perennial subshrub	Flowering times :	June - October

Flowers : small pink (sometimes white) bell-shaped drooping flowers borne in compact clusters at the ends of its shoots.

Leaves : linear leaves are usually glandular and in whorls of four.

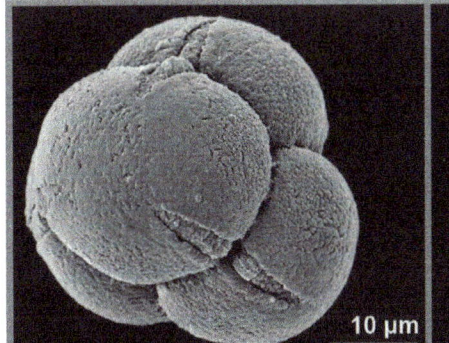 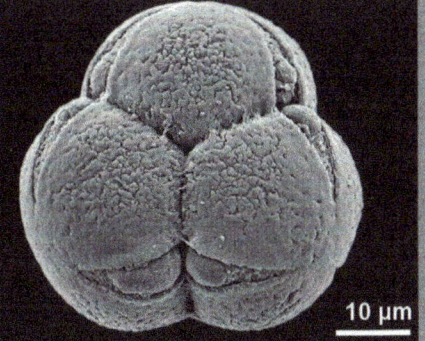 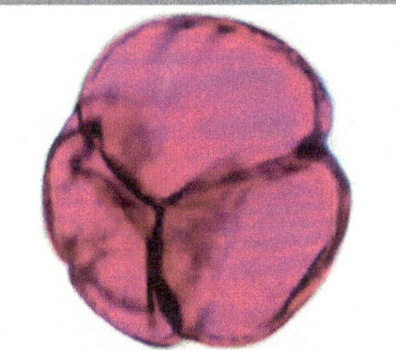

Pollen : Pollen loads are white. The tetrads are 40 um in diameter. The pollen grains are grouped together in groups of 4 in a tetrahedral configuration called 'tetrads'. The pollen grains are irregularly round with 3 aperutres which are furrows with pores. The surface structure is smooth or indefinite and the exine is of meduim thickness. There appears to be thickened or projecting edges to apertures.

Nectar presentation : No honey crop.

Preferred habitat : Acid bogs, wet heaths, moorland.

Particular special features : Having a longer flower tube (7-8 mm) this would not be a honey bee plant. However, short-tongued bumblebees puncture the base of the corolla to steal the nectar which also allows honey bees access.

It has a Pollen Coefficient of 10 which means it is 'under-represented' in honey.

SEM pictures by Halbritter, H. In: PalDat - A palynological database.
https://www.paldat.org/pub/Erica_tetralix/304301; accessed 2022-03-31 Polar and Equatorial

Winter heath

Family :	Ericaceae	Latin Name :	*Erica carnea*
Common Name :	Winter Heath	Alternate Names :	Winter flowering heather, spring heath
Description :	Low-growing, spreading subshrub	Flowering times :	December - March

Flowers : are produced in racemes. The individual flower is a slender bell-shape, 4-6 mm long, dark reddish-pink, rarely white..

Leaves : with evergreen needle-like leaves 4-8 mm long, borne in whorls of four.

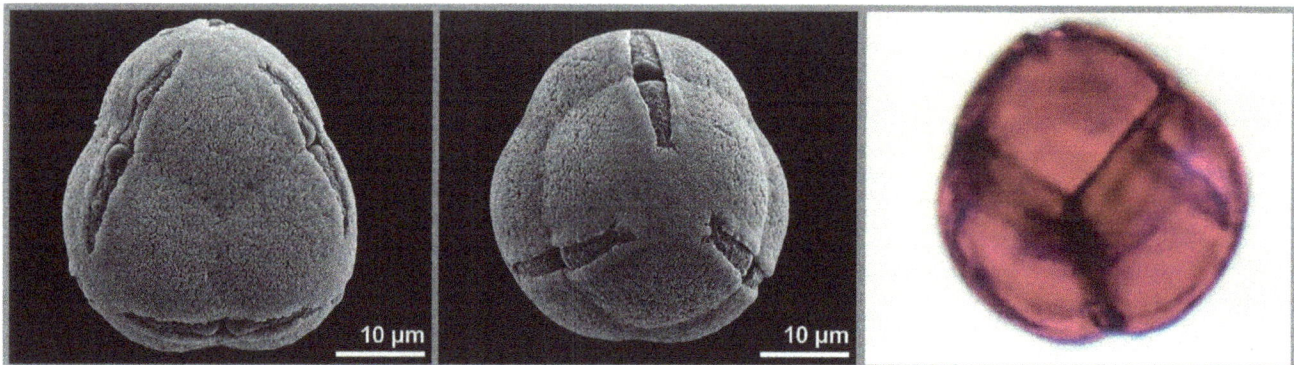

Pollen : Pollen loads are white-grey. The tetrads are 30 um in diameter. The pollen grains are grouped together in groups of 4 in a tetrahedral configuration called 'tetrads'. The pollen grains are irregularly round with 3 apertures which are furrows with pores. The surface structure is smooth or indefinite and the exine is of meduim thickness. There appears to be thickened or projecting edges to apertures.

Nectar presentation : Not a significant crop. Light to dark yellow colour, strong in flavour and aroma.

Preferred habitat : Dry heathland, open woodland, maritime heaths on acid well-drained soils.

Particular special features : Good early nectar plants for honey bees, bumblebees and solitary bees.

It has a Pollen Coefficient of 10 which means it is 'under-represented' in honey.

SEM pictures by Halbritter, H. In: PalDat - A palynological database.
https://www.paldat.org/pub/Erica_carnea/304303; accessed 2022-03-31 Polar and Equatorial

Darley Dale Heath

Family : Ericaceae
Common Name : Darley Dale Heath

Latin Name : *Erica darleyensis*
Alternate Name : None

Description : Low-growing, spreading subshrub

Flowering times : December - May

Flowers : clusters of lilac-pink, urn-shaped flowers smother this plant.

Leaves : the small, lance-shaped, leaves are mid-green, with pink tips in spring.

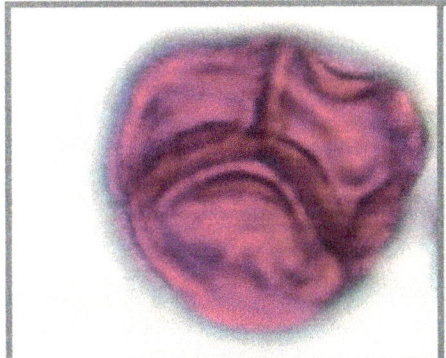
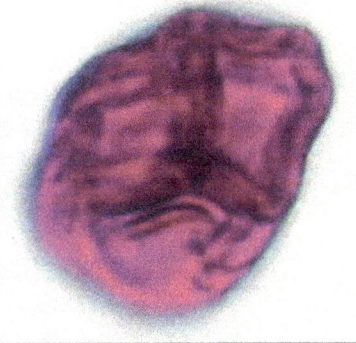
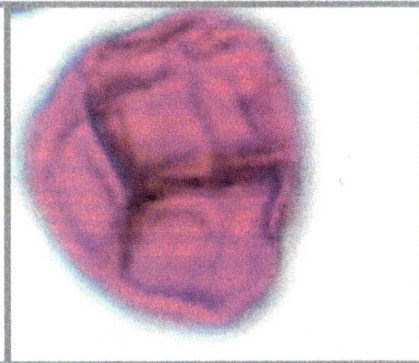

Pollen : Pollen loads are yellow. The tetrads are 30 um in diameter. The pollen grains are grouped together in groups of 4 in a tetrahedral configuration called 'tetrads'. The pollen grains are irregularly round with 3 aperutres which are furrows with pores. The surface structure is smooth or indefinite and the exine is of meduim thickness. There appears to be thickened or projecting edges to apertures.

Nectar presentation : No honey crop.

Preferred habitat : Dry heathland, open woodland, maritime heaths on acid well-drained soils.

Particular special features : Good early nectar plants for honey bees, bumblebees and solitary bees.

It has a Pollen Coefficient of 10 which means it is 'under-represented' in honey.

Rhododendron

Family : Ericaceae
Common Name : Rhododendron

Latin Name : *Rhododendron ponticum*
Alternate Names : None

Description : Dense, suckering shrub or small tree
Flowering times : May - June

Flowers : are 3.5 to 5 cm in diameter, violet-purple, often with small greenish-yellow spots or streaks.

Leaves : tough, leathery, dark green, oval leaves from 6 - 12 cm long.

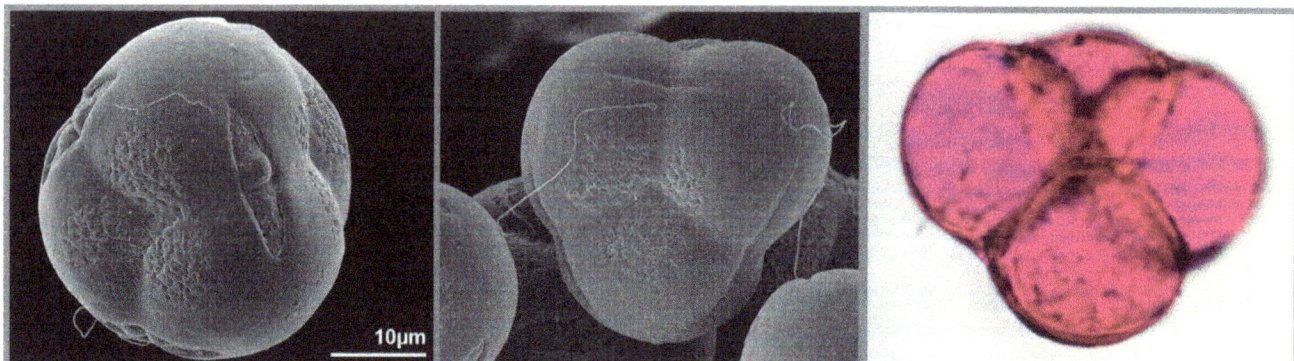

Pollen : Pollen loads are white-grey. The tetrads are 30 um in diameter. The pollen grains are grouped together in groups of 4 in a tetrahedral configuration called 'tetrads'. The pollen grains are irregularly round with 3 apertures which are furrows with pores. The surface structure is smooth or indefinite and the exine is of meduim thickness. There appears to be thickened or projecting edges to apertures.

Nectar presentation : No honey crop - toxic to honey bees. The nectar's grayanotoxins cause palpitations, paralysis and death within hours for honey bees. However, it seems that the Caucasian subspecies (Apis mellifera caucasica) is able to work the flowers and be unaffected. The honey that results is said to be intoxicating to humans. There are Greek folk tales that decsribe the use of the honey to put invading armies at a disadvantage. For the more adventurous reader it is known as 'Mad Honey' or 'Deli Bal'. Do please let me know how you get on.

Preferred habitat : Woods, heaths, rocky hillsides on acid soils.

Particular special features : This is a good bumblebee plant for both short and long-tongued species. It is an invasive non-native species and so should not be cultivated.

Note - I have selected the most common 'escape' Rhododendron. There are hundreds of others and Azaleas which are similarly toxic to bees.

SEM pictures by Halbritter, H. In: PalDat - A palynological database.
https://www.paldat.org/pub/Rhododendron_ferrugineum/304294; accessed 2022-03-31 Polar and Multiple grains

Calico bush

Family : Ericaceae
Common Name : Calico bush

Latin Name : *Kalmia latifolia*
Alternate Name : Mountain laurel or Spoonwood

Description : A broadleaved evergreen shrub

Flowering times : May - August

Flowers : are round, ranging from light pink to white, and occur in clusters. To me they look like little umbrellas.

Leaves : are 3 - 12 cm long and 1 - 4 cm wide.

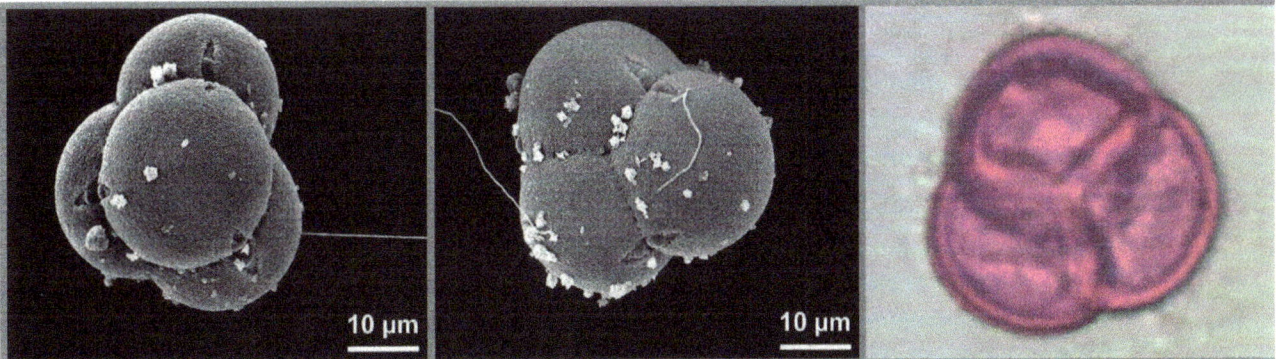

Pollen : Pollen loads are yellow. The tetrads are 40 um in diameter. The pollen grains are grouped together in groups of 4 in a tetrahedral configuration called 'tetrads'. The pollen grains are irregularly round with 3 apertures which are furrows with pores. The surface structure is smooth or indefinite and the exine is of meduim thickness. There appears to be thickened or projecting edges to apertures.

Nectar presentation : No honey crop - poisonous to honey bees and humans due to grayanotoxins.

Preferred habitat : Exposed ridges, plateaux, moorland on acid soil.

Particular special features : Bumblebees and solitary bees collect nectar from this plant.

SEM pictures by Halbritter, H. In: PalDat - A palynological database.
https://www.paldat.org/pub/Kalmia_latifolia/302153; accessed 2022-03-31 Polar and Equatorial

Blueberry

Family :	Ericaceae	Latin Name :	*Vaccinium corymbosum*
Common Name :	Blueberry	Alternate Names :	Blue huckleberry, tall huckleberry, swamp huckleberry, high blueberry, and swamp blueberry.
Description :	Food crop of economic importance.	Flowering times :	May - June

Flowers : are long bell or urn-shaped white to very light pink, 8.4 mm long.

Leaves : The dark glossy green leaves are elliptical and up to 5 cm long.

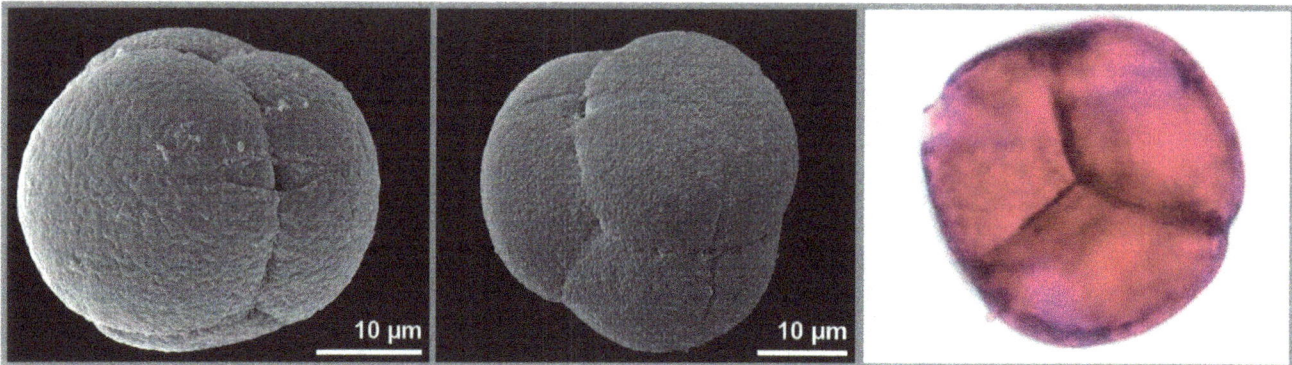

Pollen : Pollen loads are pink. The tetrads are 30 um in diameter. The pollen grains are grouped together in groups of 4 in a tetrahedral configuration called 'tetrads'. The pollen grains are irregularly round with 3 apertures which are furrows with pores. The surface structure is smooth or indefinite and the exine is of meduim thickness. There appears to be thickened or projecting edges to apertures.

Nectar presentation : No honey crop.

Preferred habitat : Moors, heathland, open acid woodlands and drier parts of peat bogs.

Particular special features : Visited by honey bees, solitary bees and bumblebees. However, commercially they are pollinated using bumblebees. Bumblebees employ 'buzz' polination where they grab the anthers and vibrate their wing muscles and the vibrations cause the anthers to release the pollen. It was thought that this was the most ecconomical way to pollinate blueberries. However, recent research has shown that while the honey bee rests on the flower and grooms herself her backlegs enter the flower and pollination occurs - seemingly by accident.

A Novel Pollen Transfer Mechanism by Honey Bee Foragers on Highbush Blueberry (Ericales: Ericaceae) - Environmental Entomology, Volume 47, Issue 6, December 2018, Pages 1465-1470.

SEM pictures by Halbritter, H. In: PalDat - A palynological database.
https://www.paldat.org/pub/Vaccinium_uliginosum/306206; accessed 2022-03-31 Polar and Equatorial

Strawberry Tree

Family :	Ericaceae	Latin Name :	*Arbutus unedo*
Common Name :	Strawberry tree	Alternate Name :	Cain or cane apple, Killarney strawberry
Description :	Evergreen shrub or small tree	Flowering times :	September - December

Flowers : are white (rarely pale pink), bell-shaped, 4-6 mm diameter, produced in panicles of 10-30.

Leaves : are dark green and glossy, 5 - 10 cm long and 2 - 3 cm broad, with a serrated margin.

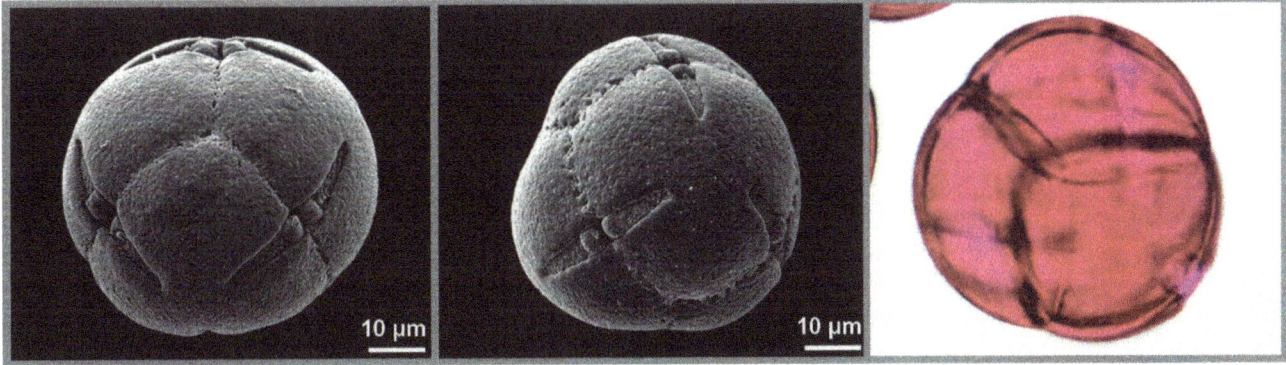

Pollen : Pollen loads are brown. The tetrads are 50 um in diameter. The pollen grains are grouped together in groups of 4 in a tetrahedral configuration called 'tetrads'. The pollen grains are irregularly round with 3 apertures which are furrows with pores. The surface structure is smooth or indefinite and the exine is of meduim thickness. There appears to be thickened or projecting edges to apertures.

Nectar presentation : No honey crop.

Preferred habitat : Scrub, open woodland and rocky lake shores.

Particular special features : A good source of autumn nectar and pollen for honey bees and short-tongued bumblebees such as buff-tails *Bombus terrestris*.

SEM pictures by Halbritter, H. In: PalDat - A palynological database.
https://www.paldat.org/pub/Arbutus_unedo/302416; accessed 2022-03-31 Polar and Equatorial

Pieris

Family :	Ericaceae	Latin Name :	*Pieris* **spp**
Common Name :	Pieris	Alternate Names :	Andromedas or Fetterbushes
Description :	Broad-leaved evergreen shrubs	Flowering times :	February - May

Flowers : are bell-shaped, 5 - 15 mm long, white or pink, and arranged in racemes 5 - 12 cm long.

Leaves : are spirally arranged, often appearing to be in whorls at the end of each shoot. They are lanceolate-ovate, 2 - 10 cm long and 1.0 - 3.5 cm broad.

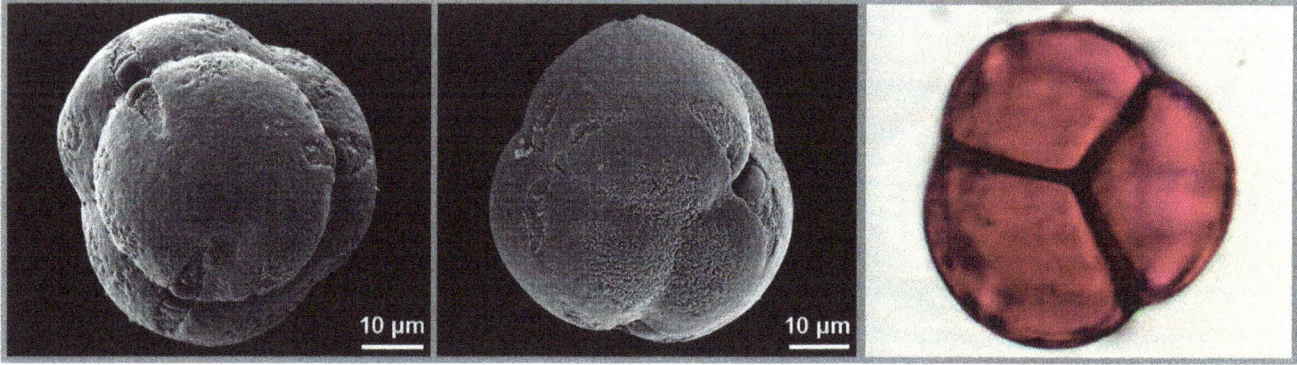

Pollen : Pollen loads are white. The tetrads are 30 um in diameter. The pollen grains are grouped together in groups of 4 in a tetrahedral configuration called 'tetrads'. The pollen grains are irregularly round with 3 apertures which are furrows with pores. The surface structure is smooth or indefinite and the exine is of meduim thickness. There appears to be thickened or projecting edges to apertures.

Nectar presentation : No honey crop.

Preferred habitat : Usually from China or the Himalayas and although preferring acid soils - it is seldomfound on Sphagnum and UK bogs but more normal on acid garden soils.

Particular special features : A good early source of nectar for bumblebees. Because of the length of the flower tube (circa 7 mm) it is doubtful that honey bees can reach the nectar.

SEM pictures by Halbritter, H. In: PalDat - A palynological database.
https://www.paldat.org/pub/Vaccinium_vitis-idaea/304304; accessed 2022-03-31 Polar and Equatorial

Bilberry

Family :	Ericaceae	Latin Name :	***Vaccinium myrtillus***
Common Name :	Bilberry	Alternate Name :	Wimbleberry, Whortleberry
Description :	A species of shrub with edible fruit	Flowering times :	April - May

Flowers : are borne singly in leaf axils on 2 - 3 mm long pedicels. The corolla is white to pink and shaped like an urn.

Leaves : are finely toothed and prominently veined on the lower surface.

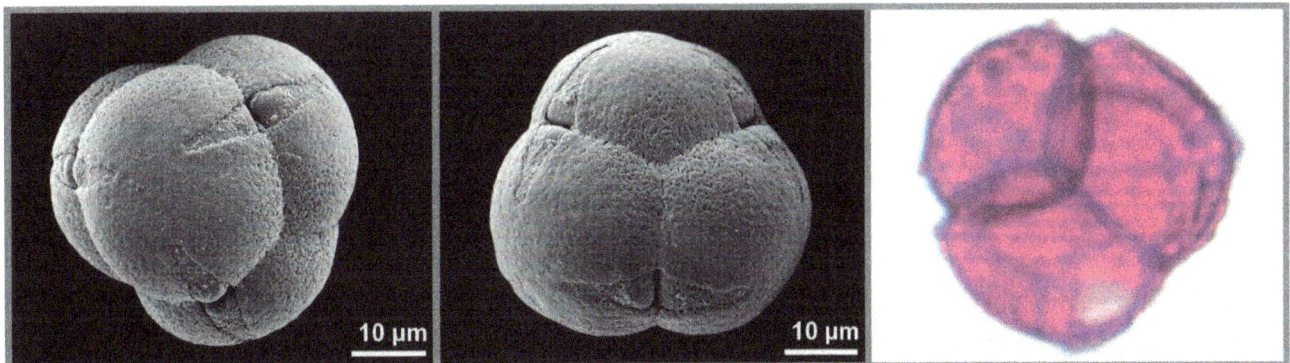

Pollen : Pollen loads are white. The tetrads are 40 um in diameter. The pollen grains are grouped together in groups of 4 in a tetrahedral configuration called 'tetrads'. The pollen grains are irregularly round with 3 apertures which are furrows with pores. The surface structure is smooth or indefinite and the exine is of meduim thickness. There appears to be thickened or projecting edges to apertures.

Nectar presentation : No honey crop.

Preferred habitat : Moors, heathland, open acid woodlands and drier parts of peat bogs.

Particular special features : A good nectar producer, it is visited by honey bees, bumblebees and solitary bees.

SEM pictures by Halbritter, H. In: PalDat - A palynological database.
https://www.paldat.org/pub/Vaccinium_myrtillus/304305; accessed 2022-03-31 Polar and Equatorial

Cranberry

Family : Ericaceae
Common Name : Cranberry

Latin Name : *Vaccinium oxycoccos*
Alternate Names : Small cranberry, Bog or Swamp cranberry

Description : A species of flowering plant

Flowering times : June - July

Flowers : arise on nodding stalks a few cm tall. The corolla is white or pink and flexed backward away from the centre of the flower.

Leaves : are leathery and lance-shaped, up to 1 cm long.

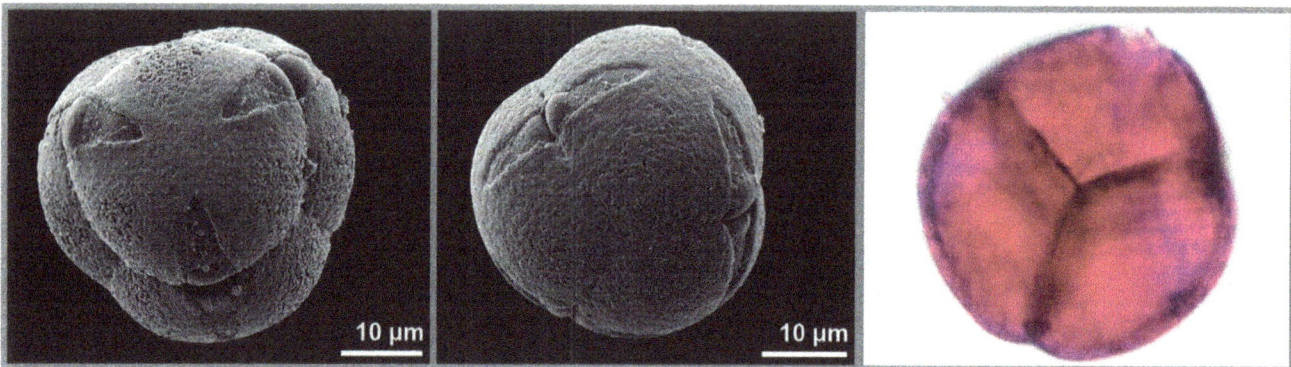

Pollen : Pollen loads are white. The tetrads are 30 um in diameter. The pollen grains are grouped together in groups of 4 in a tetrahedral configuration called 'tetrads'. The pollen grains are irregularly round with 3 apertures which are furrows with pores. The surface structure is smooth or indefinite and the exine is of meduim thickness. There appears to be thickened or projecting edges to apertures.

Nectar presentation : No honey crop.

Preferred habitat : Wet acid bogs and heaths, open woodland often creeping over sphagnum moss.

Particular special features : Commercially pollinated by both honey bees and bumblebees. Cranberries are mostly grown commercially in the USA and not the UK (although I believe there is one farm in Kent)

SEM pictures by Halbritter, H. In: PalDat - A palynological database.
https://www.paldat.org/pub/Vaccinium_vitis-idaea/304304; accessed 2022-03-31 Polar and Equatorial

Boraginaceae - Borage Family

Key words : Hairy plants with flower parts in fives

Boraginaceae, the borage or forget-me-not family, includes a variety of shrubs, trees, and herbs, totaling about 2,000 species in 146 genera found worldwide.

Most pollination is by hymenopterans. Most species have inflorescences that have a coiling shape, at least when new.
Plants are often rough or hairy with simple alternate leaves.

The flower spikes often curve with the flowers blooming on the upper surface.
Individual flowers are bisexual. They have 5 separate sepals and 5 fused petals. There are 5 stamens alternate with the petals.
It has a superior ovary with 2 fused carpels, although it appears as 4 chambers. It typically matures as 4 separate nutlets.

Boraginaceae are known to sometimes contain pyrrolizidine alkaloids (PAs) - see page 112

Species of importance to bees :

1. Common Name : Borage
   Latin Name :   **Borago officinalis**

2. Common Name : Forget-me-not
   Latin Name :   **Myosotis spp**

3. Common Name : Vipers Bugloss
   Latin Name :   **Echium vulgare**

4. Common Name : Phacelia
   Latin Name :   **Phacelia tanacetifolia**

5. Common Name : Common Lungwort
   Latin Name :   **Pulmonaria officinalis**

6. Common Name : Comfrey
   Latin Name :   **Symphytum spp**

Osmia bicornis - male Red Mason Bee foraging on forget-me-not

Honey bee foraging on Borage

Borage

Family :	Boraginaceae	Latin Name :	**Borago officinalis**
Common Name :	Borage	Alternate Names :	Starflower
Description :	An annual herb	Flowering times :	June - September

Flowers : are most often blue, although pink flowers are sometimes observed. The flowers are complete, perfect with five narrow, triangular-pointed petals.

Leaves : are alternate, simple, and 5 - 15 cm long. The leaves are edible and the plant is grown in gardens for that purpose.

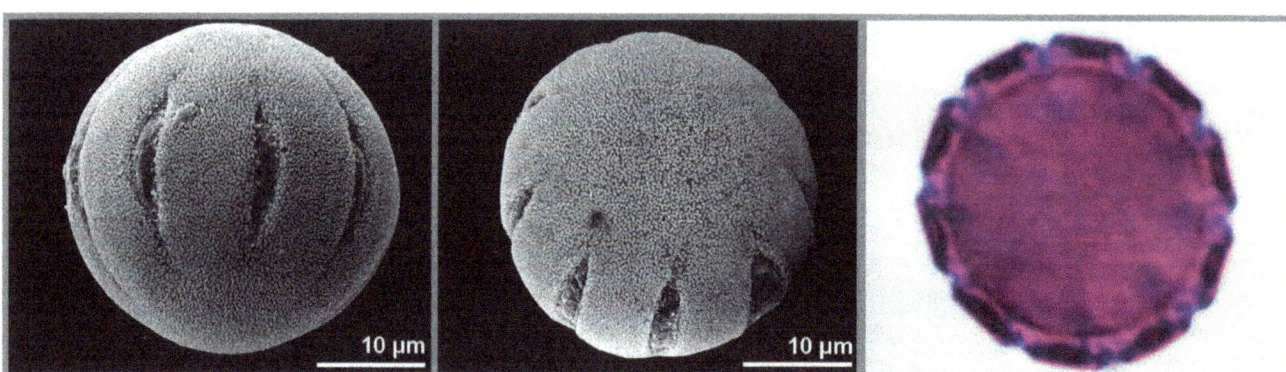

Pollen : Pollen loads are white. The pollen grains are 30 um in diameter. The pollen grains are round with 7 - 12 apertures which are furrows . The surface structure is smooth or indefinite and the exine is of medium thickness. There are no apparent surface features on the apertures.

Nectar presentation : Not considered a major honey crop - colour is very light - whitish with a yellow grey tint. A very sweet honey lacking in flavour.

Preferred habitat : Roadsides and waste ground.

Particular special features : Grown commercially for the oil, they are great bee plants. The drooping nature of the flowers means that they are not washed-out by rain. They secrete nectar profusely over a long flowering season. Visited by honey bees, solitary bees and bumblebees, the later employing the 'buzz' technique for obtaining the pollen.

It is not listed in Sawyer's Honey Identification but it is under-represented in honey. Finding borage pollen grains in honey means the honey is largely from borage.

SEM pictures by Halbritter, H. In: PalDat - A palynological database.
https://www.paldat.org/pub/Borago_officinalis/304382; accessed 2022-03-31 Polar and Equatorial

Forget-me-not

Family : Boraginaceae
Common Name : Forget-me-not

Latin Name : **_Myosotis_ spp**
Alternate Name : Scorpion grasses or Mouse's ear

Description : Annual or perennial flowering plants
Flowering times : April - September

Flowers : are typically 1 cm diameter (or less), flat, and blue, pink, white or yellow with yellow centres, growing on scorpioid cymes. (A type of cyme in which the axis is curved or coiled like the tail of a scorpion and the flowers open successively downwards)

Leaves : alternate, simple, and entire.

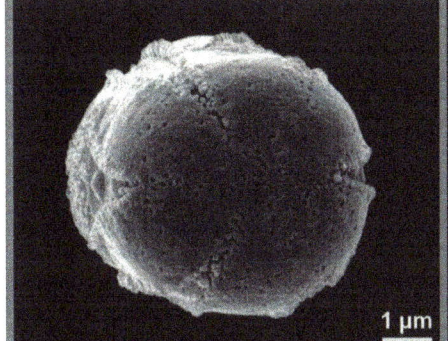

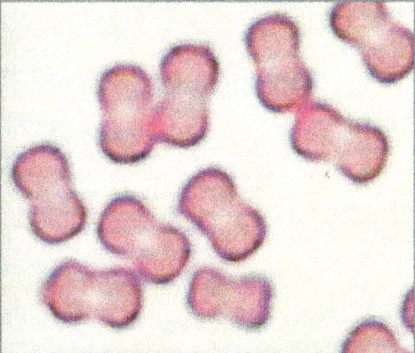

Pollen : Pollen loads are yellow - orange. The pollen grains are 2 x 5 um. The pollen grains are long with 7 - 12 apertures which are furrows with pores. The surface structure is smooth or indefinite and the exine is of medium thickness. There are no surface features on the apertures.

Nectar presentation : No honey crop.

Preferred habitat : Roadsides, woodland, grassy heaths and disturbed ground on arable soils.

Particular special features : A good source of early nectar and pollen. It is the smallest of the pollen grains and as such is often found in honey. It has a Pollen Coefficient of 5000 which means it is 'over-represented' in honey. The reason for this is the position of the anthers directly above the nectaries at the top of the corolla. As the honey bee pushes past the anthers to reach the nectaries pollen is dropped into the nectar. Being so small it can be carried in the nectar to the crop or honey stomach of the bee.

Efficient pollination is helped by a colour change in the coronal ring of the flower. The flowers start off with a yellow ring in the centre of the flower. When the flower is pollinated this ring turns white and essentially the flower is 'turned off' to pollinating insects.

SEM pictures by Halbritter, H. In: PalDat - A palynological database.
https://www.paldat.org/pub/Myosotis_arvensis/304389; accessed 2022-03-31 Polar and Equatorial

Viper's Bugloss

Family : Boraginaceae
Common Name : Viper's Bugloss
Description : A biennial or monocarpic perennial.

Latin Name : *Echium vulgare*
Alternate Names : Blueweed
Flowering times : May - September

Flowers : the flowers start pink and turn vivid blue and are 15 - 20 mm in a branched spike, with all the red stamens protruding.

Leaves : rough, hairy, oblanceolate (a leaf having a rounded apex and tapering base) leaves.

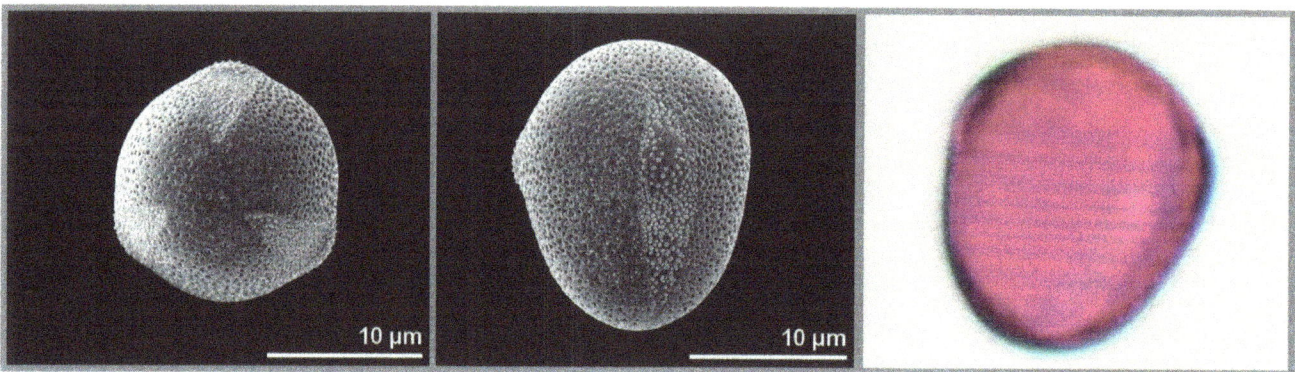

Pollen : Pollen loads are dark brown - black. The pollen grains are 10 x 15 um. The pollen grains are oval elongated with 3 apertures which are furrows with pores. The surface is smooth or indefinite and the exine is thin. There are no surface features on the apertures.

Nectar presentation : No honey crop.

Preferred habitat : Open disturbed ground, rough grassland, cliffs, dunes, shingle, waste ground on light calcareous soils.

Particular special features : An excellent bee plant, it is often included in wild flower seed mixes. It is visited by honey bees, bumblebees and solitary bees and secretes nectar freely.

It has a Pollen Coefficient of 250 which means it is 'over-represented' in honey.

SEM pictures by Halbritter, H. In: PalDat - A palynological database.
https://www.paldat.org/pub/Echium_vulgare/304395; accessed 2022-03-31 Polar and Equatorial

Phacelia

Family :	Boraginaceae	Latin Name :	*Phacelia tanacetifolia*
Common Name :	Phacelia	Alternate Name :	Lacy phacelia, Blue tansy or Purple tansy
Description :	An annual herb which grows erect	Flowering times :	March - September

Flowers : a one-sided curving or coiling cyme of bell-shaped flowers in shades of blue and lavender. Each flower is just under 1 cm long and has protruding whiskery stamens.

Leaves : are mostly divided into smaller leaflets deeply and intricately cut into toothed lobes, giving them a lacy appearance.

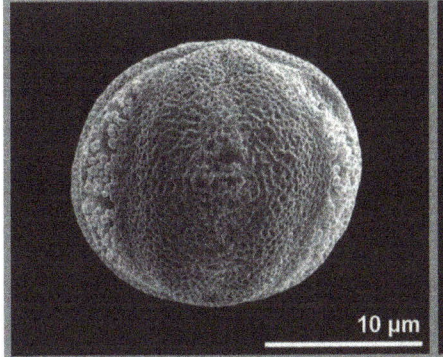
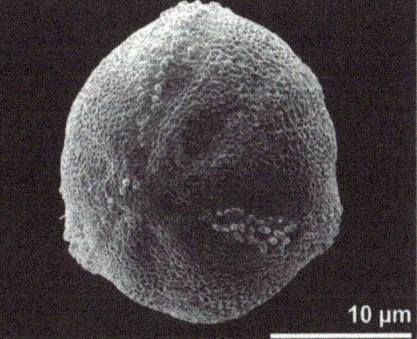
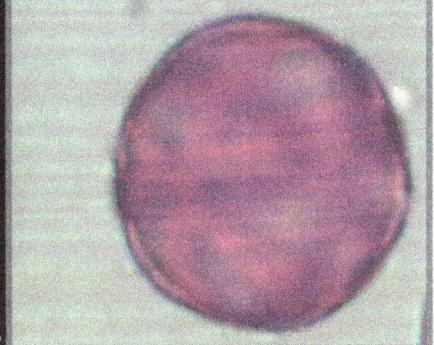

Pollen : Pollen loads are dark blue. The pollen grains are 20 um diameter. The pollen grains are round with 4-6 apertures which are furrows. The surface structure is smooth or indefinite and th exine is thin. There are granules or projections on the apertures.

Nectar presentation : Not considered a major honey crop - colour is white to amber. Flavour mild. Granulation rapid.

Preferred habitat : Cultivated ground and rubbish tips.

Particular special features : It is a very good honey bee plant and is also visited by short-tongued bumblebees.

SEM pictures by Halbritter, H. In: PalDat - A palynological database.
https://www.paldat.org/pub/Phacelia_tanacetifolia/300737; accessed 2022-03-31 Polar and Equatorial

Common Lungwort

Family :	Boraginaceae	Latin Name :	***Pulmonaria officinalis***
Common Name :	Common Lungwort	Alternate Names :	Mary's tears or Our Lady's milk drops
Description :	A rhizomatous evergreen perennial	Flowering times :	March - May

Flowers : the 5-petalled flowers are red or pink at first, later turn to blue-purple

Leaves : basal leaves are green, cordate, more or less elongated and pointed and always with rounded and often sharply defined white or pale green patches.

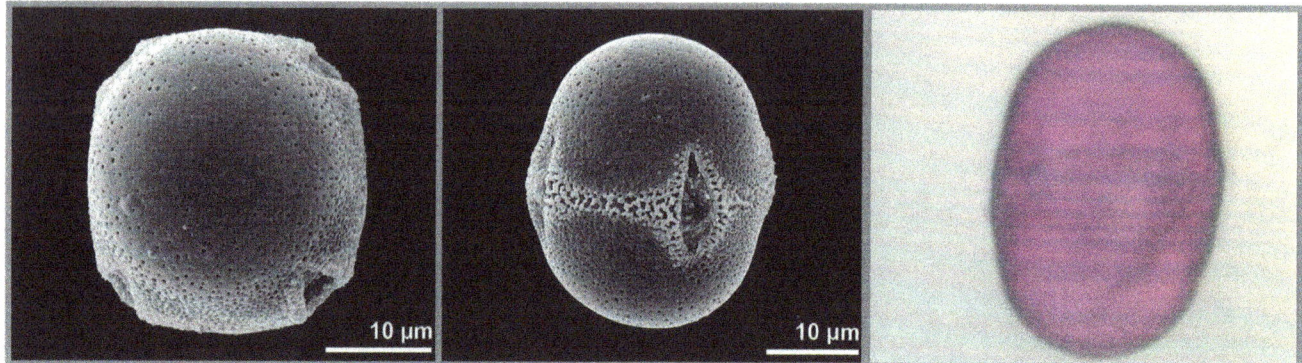

Pollen : Pollen loads are white. The pollen grains are 30 um in diameter. The pollen grains are oval but sqaure in cross-section. There are 3 apertures which are furrows with pores. The surface structure is smooth or indefinite and the exine appears to be of medium thickness. There are no surface features on the apertures.

Nectar presentation : No honey crop.

Preferred habitat : Woods, hedgebanks and rough ground on not too-dry chalky soils.

Particular special features : The flower tube or corolla is quite long and so prevents honey bees and short-tongued bumblebees from accessing the nectar. However, they may well access the pollen.

SEM pictures by Halbritter, H. In: PalDat - A palynological database.
https://www.paldat.org/pub/Pulmonaria_officinalis/304385; accessed 2022-03-31 Polar and Equatorial

Comfrey

Family :	Boraginaceae	Latin Name :	*Symphytum* spp
Common Name :	Comfrey	Alternate Name :	Comphrey
Description :	Important herb in organic gardening	Flowering times :	May - July

Flowers : small bell-shaped flowers of various colours, typically cream or purplish, which may be striped.

Leaves : large, hairy broad leaves

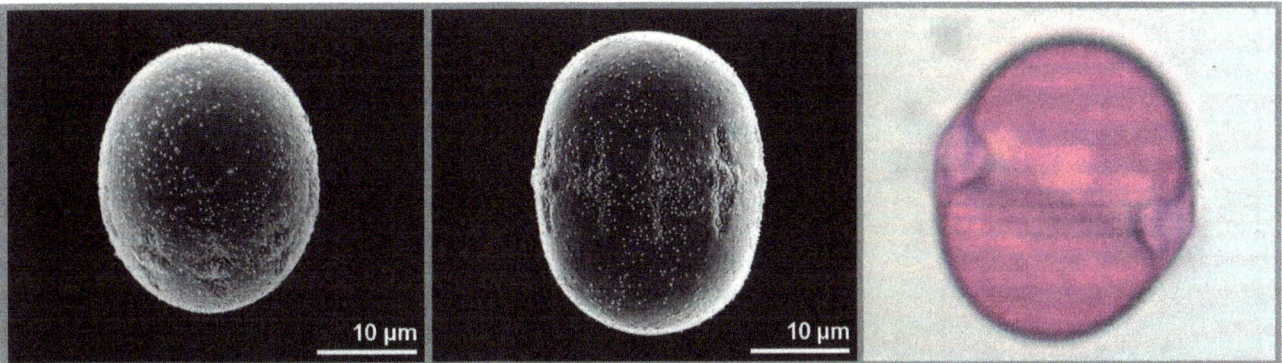

Pollen : Pollen loads are white - yellow. The pollen grains are 20 x 30 um. The pollen grains are oval elongated with 7 - 12 apertures which are furrows with pores. The surface structure is smooth or indefinite and the exine is thin. There are no surface features on the apertures.

Nectar presentation : No honey crop.

Preferred habitat : Fens, marshes, wet ditches, banks of rivers, streams and canals.

Particular special features : This is another flower with a long corolla or flower tube. This forces short-tongued bumblebees to pierce a hole in the base of the tube to access the netar. Honey bees make use of thes holes to access the nectar too. This is also another flower species to which the bumblebees use the 'buzz' technique to extract the pollen efficiently.

Very attractive to long-tongued bumblebees and bee flies.

SEM pictures by Bombosi, P. In: PalDat - A palynological database.
https://www.paldat.org/pub/Symphytum_officinale/304380; accessed 2022-03-31 Polar and Equatorial

Oleaceae - Olive Family

Key words : Opposite leaves, 4 sepals, 4 petals and 2 stamens

The Oleaceae are a family of flowering plants in the order Lamiales. It presently comprises 26 genera.

The number of species in the Oleaceae is variously estimated in a wide range around 700. The Oleaceae consist of shrubs, trees, and a few lianas (climbing vines).

The flowers are often numerous and highly odoriferous. The family has a distribution, ranging from the subarctic to the southernmost parts of Africa, Australia, and South America.

Notable members of the Oleaceae include olive, ash, jasmine, and several popular ornamental plants including privet, forsythia, fringetrees, and lilac.

The leaves are usually opposite sometimes on squarish stems.

The flowers are regular, bisexual and often aromatic.
There are typically 4 untied sepals and 4 united petals and usually 2 stamens.

The ovary is superior and consists of 2 unted carpels that form 2 chambers.
It can mature as a capsule, a pair of winged seeds, a fleshy fruit with a stone pit (like olive) or rarely as a berry.

Species of importance to bees :

1. Common Name : Privet
   Latin Name : **Ligustrum vulgare**

Honey bee foraging on privet in Pershore in 2020

Privet hedge in flower

Privet

Family :	Oleaceae	Latin Name :	*Ligustrum vulgare*
Common Name :	Privet	Alternate Names :	Wild privet, Common or European privet
Description :	Common hedging shrub	Flowering times :	June - August

Flowers : in panicles 3 - 6 cm long, each flower creamy-white, with a tubular base and a four-lobed corolla 'petals' 4 - 6 mm diameter. Produce a strong, pungent fragrance that many people find unpleasant.

Leaves : are borne in opposite pairs, sub-shiny green, narrow oval to lanceolate, 2 - 6 cm long and 0.5 - 1.5 cm broad.

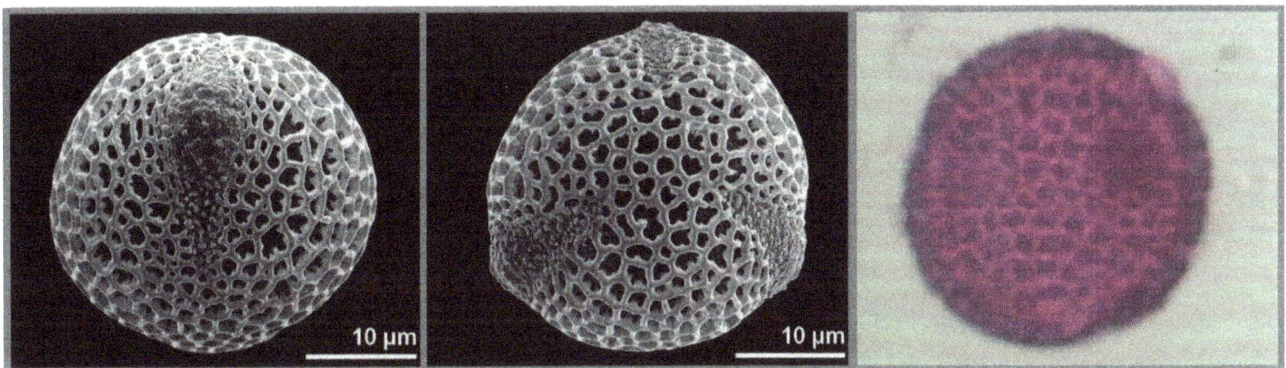

Pollen : Pollen loads are yellow. The pollen grains are 30 um in diameter. The pollen grains are round with 3 apertures which are furrows with pores. The surface structure is netted or pitted with a thick exine that has course external rods. The intine appears swollen beneath the apertures.

Nectar presentation : No honey crop. The resulting honey is 'unpalatable'. It has a strong bitter flavour.

Preferred habitat : Old hedgerows, scrub, wood borders on well-drained calcareous soils.

Particular special features : The flowers have an unpleasant odour but are very attractive to honey bees, bumble bees and solitary bees.

It has a Pollen Coefficient of 25 which means it is 'normally represented' in honey.

SEM pictures by Halbritter, H. In: PalDat - A palynological database.
https://www.paldat.org/pub/Ligustrum_vulgare/306267; accessed 2022-03-31 Polar and Equatorial

Lamiaceae - Mint or Deadnettle Family

Key words : Usually aromatic with square stalks and opposite leaves

The Lamiaceae are a family of flowering plants commonly known as the mint or deadnettle family.

Many of the plants are aromatic in all parts and include widely used culinary herbs.

Some species are shrubs, trees (such as teak), or rarely, vines.

Many members of the family are widely cultivated, not only for their aromatic qualities, but also their ease of cultivation, since they are readily propagated by stem cuttings.

Identification of the family begins with square stems with simple and opposite leaves, It the plant is aromatic then it is definitely this family.

The flowers are bisexual and irregular. They have 5 united sepals and 5 united petals. Typically with 2 lobes up and 3 lobes down.
Inside the flower are 4 stamens with 1 pair longer than the other.

The ovary is superior and consists of 2 united carpels. It matures as a capsule containing 4 nutlets.

Lamiaceae are known to sometimes contain pyrrolizidine alkaloids (PAs) - see page 112

Species of importance to bees :

1. Common Name : Thyme
   Latin Name :   *Thymus* spp

2. Common Name : Mint
   Latin Name :   *Mentha* spp

3. Common Name : Lavender
   Latin Name :   *Lavandula angustifolia*

4. Common Name : Rosemary
   Latin Name :   *Rosmarinus officinalis*

5. Common Name : Phlomis
   Latin Name :   *Phlomis*

6. Common Name : Self-heal
   Latin Name :   *Prunella grandiflora*

7. Common Name : Hyssop
   Latin Name :   *Hyssopus officinalis*

8. Common Name : Marjoram
   Latin Name :   *Origanum majorana*

9. Common Name : Catmint
   Latin Name :   *Nepeta* spp

10. Common Name : Sage
    Latin Name :   *Salvia officinalis*

11. Common Name : Lemon balm
    Latin Name :   *Melissa officinalis*

Tree bumblebee *Bombus hypnorum* foraging on phlomis

Red-tailed bumblebee *Bombus lapidarius* foraging on thyme

Thyme

Family : Lamiaceae
Common Name : Thyme

Latin Name : *Thymus* spp
Alternate Names : None
Flowering times : May - August

Description : 350 species of aromatic perennials

Flowers : are in dense terminal heads, with an uneven calyx, with the upper lip three-lobed, yellow, white or purple.

Leaves : are evergreen in most species, arranged in opposite pairs, oval, entire, and small, 4 - 20 mm long, and usually aromatic.

Pollen : Pollen loads are yellow - orange. The pollen grains are 35 um in diameter. The pollen grains are oval-irregular in shape with 6 apertures which are furrows. The surface structure is netted or pitted and the exine is of medium thickness. The surface of the apertures has scattered granules or projections.

Nectar presentation : A minor honey source - colour amber with a strong fragrant aroma and a strong minty flavour which does not fade on storage. Slow granulation.

Preferred habitat : Close grazed premanent grassland, maritime and mountain heaths, cliffs, limestone pavement, mature sand dunes on dry calcareous or acid soils.

Particular special features : Very good nectar producing plants over a long flowering period. Visited by honey bees, bumblebees and solitary bees.

It has a Pollen Coefficient of 5 which means it is 'under-represented' in honey.

SEM pictures by Ulrich, S. In: PalDat - A palynological database.
https://www.paldat.org/pub/Thymus_odoratissimus/300493; accessed 2022-03-31 Polar and Equatorial

Mint

Family :	Lamiaceae	Latin Name :	*Mentha* **spp**
Common Name :	Mint	Alternate Name :	None
Description :	Aromatic mostly perennial herbs	Flowering times :	June - September

Flowers : are white to purple and produced in false whorls called verticillasters.

Leaves : are arranged in opposite pairs, from oblong to lanceolate, often downy, and with a serrated margin. Leaf colors range from dark green and gray-green to purple, blue, and sometimes pale yellow.

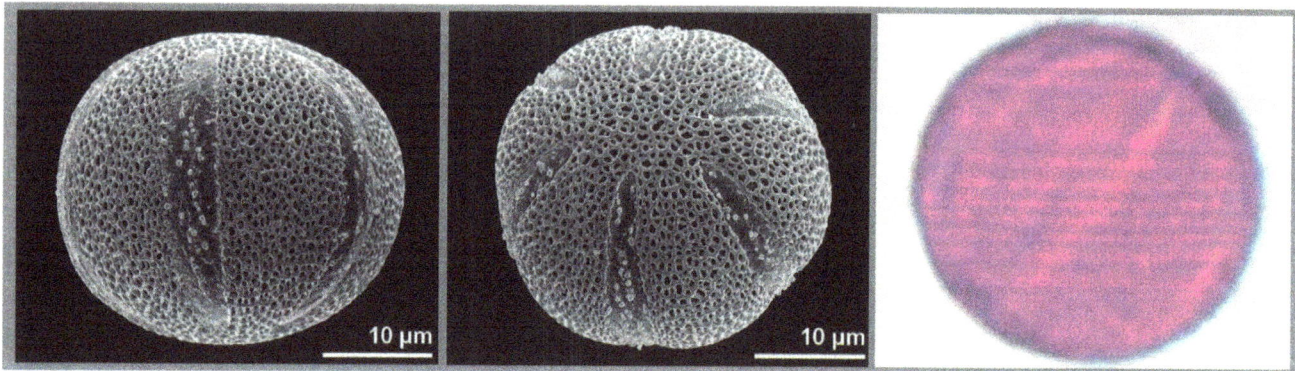

Pollen : Pollen loads are light brown. The pollen grains are 30 um in diameter. The pollen grains are oval-irregular in shape with 6 apertures which are furrows. The surface structure is netted or pitted and the exine is of medium thickness. The surface of the apertures has scattered granules or projections.

Nectar presentation : No honey crop.

Preferred habitat : Damp roadsides and waste ground.

Particular special features : All forms of mint are good honey bee plants although they are normally harvested before flowering occurs.

SEM pictures by Halbritter, H. In: PalDat - A palynological database.
https://www.paldat.org/pub/Mentha_longifolia/304454; accessed 2022-03-31 Polar and Equatorial

Lavender

Family :	Lamiaceae	Latin Name :	*Lavandula angustifolia*
Common Name :	Lavender	Alternate Names :	True lavender or English lavender
Description :	Aromatic shrub growing to 2m	Flowering times :	June - August

Flowers : flowers are pinkish-purple (lavender-coloured), produced on spikes 2-8 cm long at the top of slender, leafless stems 10-30 cm long.

Leaves : are evergreen, 2-6 cm long, and 4-6 mm broad.

Pollen : Pollen loads are orange. The pollen grains are 10 um in diameter. The pollen grains are oval-irregular in shape with 6 apertures which are furrows. The surface structure is netted or pitted and the exine is of medium thickness. The surface of the apertures has scattered granules or projections.

Nectar presentation : Not considered a major honey crop - Light in colour, strong fragrant aroma and fine granulating.

Preferred habitat : Rocky soils and in grassy, hilly meadows. It thrives in sandy, poor soils and tolerates windswept areas and even salt air.

Particular special features : It produces nectar freely and is attractive to honey bees, solitary bees and bumblebees.

SEM pictures by Halbritter, H. In: PalDat - A palynological database.
https://www.paldat.org/pub/Lavandula_angustifolia/304412; accessed 2022-03-31 Polar and Exine surface

Rosemary

Family : Lamiaceae
Common Name : Rosemary

Latin Name : *Rosmarinus officinalis*
Alternate Name : Anthos

Description : A perennial fragrant, evergreen herb
Flowering times : May - June

Flowers : flowers are white, pink, purple or deep blue. Rosemary also has a tendency to flower outside its normal flowering season

Leaves : are evergreen, 2 - 4 cm long and 2 - 5 mm broad, green above, and white below, with dense, woolly hair.

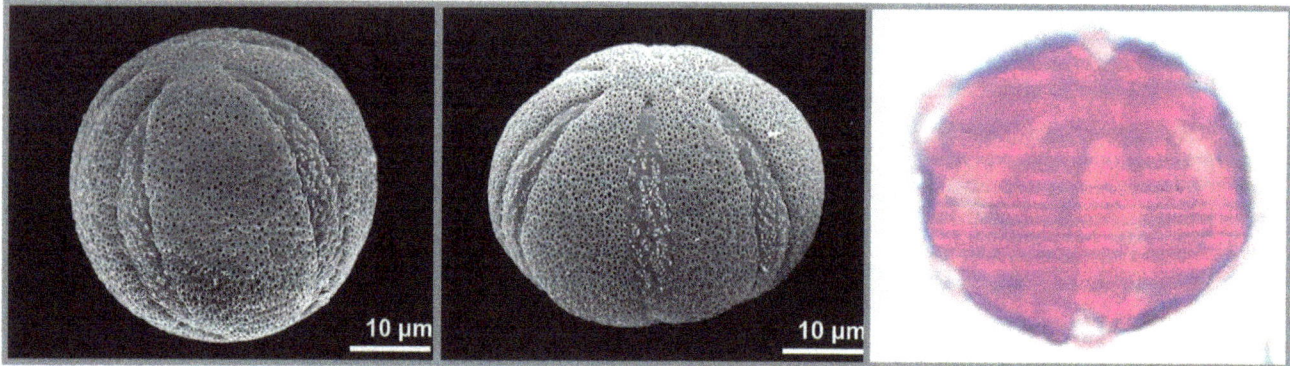

Pollen : Pollen loads are white. The pollen grains are 40 um in diameter. The pollen grains are oval-irregular in shape with 6 apertures which are furrows. The surface structure is netted or pitted and the exine is of medium thickness. The surface of the apertures has scattered granules or projections.

Nectar presentation : Not a major honey crop in the UK. Light and clear, flavour and aroma distinctive. Usually rapid, fine granulation.

Preferred habitat : Dry valleys and coastal regions with poor, dry, sandy, and rocky soil types. It is drought resistant.

Particular special features : Very attractive to honey bees, solitary bees and bumblebees.

SEM pictures by Halbritter, H. In: PalDat - A palynological database.
https://www.paldat.org/pub/Rosmarinus_officinalis/304409; accessed 2022-03-31 Polar and Equatorial

Phlomis

Family : Lamiaceae
Common Name : Phlomis

Latin Name : *Phlomis fruticosa*
Alternate Names : Jerusalem sage and lampwick plant

Description : Herbaceous perennials

Flowering times : June - July

Flowers : are arranged in whorls called verticillasters which encircle the stems. The colour of the flowers varies from yellow to pink, purple and white.

Leaves : leaves are entire, opposite and decussate (each leaf pair at right angles to the next) and rugose or reticulate veined.

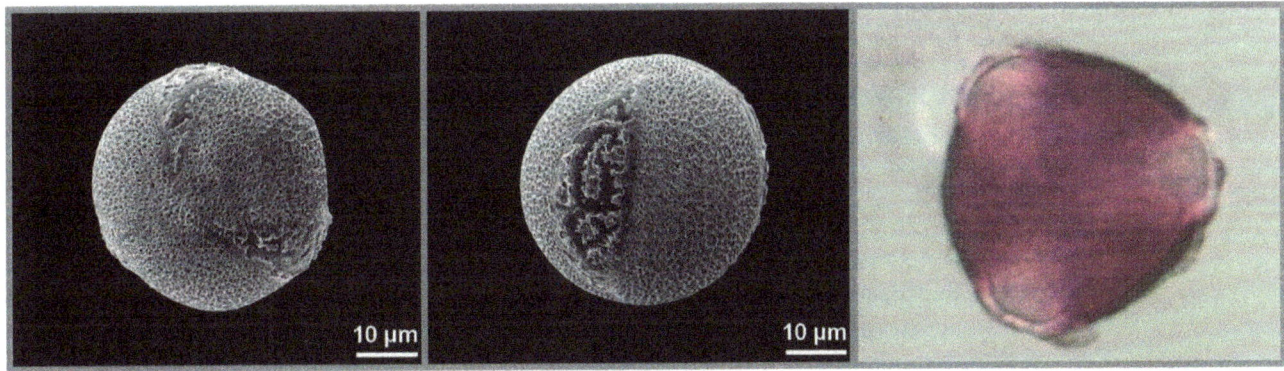

Pollen : Pollen loads are yellow. The pollen grains are 40 um in diameter. The pollen grains are round but triangular in cross-section. There are 3 apertures which are furrows with pores. The surface structure is netted or pitted and the exine is of medium thickness. There are granules or projections scattred on the apertures.

Nectar presentation : No honey crop.

Preferred habitat : Coastal areas and waste ground in moist but well-drained soil in full sun to partial shade.

Particular special features : Attractive to honey bees, solitary bees and bumblebees.

SEM pictures by Halbritter, H. In: PalDat - A palynological database.
https://www.paldat.org/pub/Phlomis_viscosa/305091; accessed 2022-03-31 Polar and Equatorial

Self-heal

Family :	Lamiaceae	Latin Name :	*Prunella grandiflora*
Common Name :	Self-heal	Alternate Name :	Large Self-heal, Large-flowered Self-heal
Description :	A pretty, perennial ground cover plant.	Flowering times :	June - July

Flowers : short spikes of hooded white to blue flowers.

Leaves : produces a thick carpet of deep green foliage, which are arranged in opposite pairs, from oblong to lanceolate.

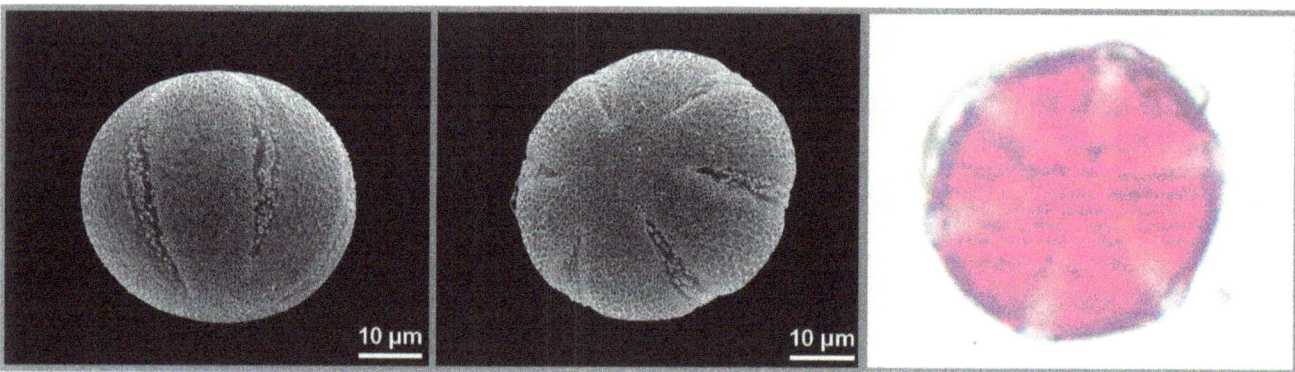

Pollen : Pollen loads are white. The pollen grains are 50 um in diameter. The pollen grains are oval-irregular in shape with 6 apertures which are furrows. The surface structure is netted or pitted and the exine is of medium thickness. The surface of the apertures has scattered granules or projections.

Nectar presentation : No honey crop.

Preferred habitat : Calcareous grasslands and dry banks.

Particular special features : Attractive to honey bees, solitary bees and bumblebees.

Note : *Prunella grandiflora* is mostly a garden plant and usually called Large-flowered self-heal to distinguish it from *Prunella vulgaris* - the native and common self-heal.

SEM pictures by Halbritter, H. In: PalDat - A palynological database.
https://www.paldat.org/pub/Prunella_grandiflora/304419; accessed 2022-03-31 Polar and Equatorial

Hyssop

Family :	Lamiaceae	Latin Name :	*Hyssopus officinalis*
Common Name :	Hyssop	Alternate Names :	None
Description :	Commonly used as a medicinal plant	Flowering times :	July - September

Flowers : bunches of pink, blue, or, more rarely, white fragrant flowers.

Leaves : are lanceolate, dark green in colour, and from 2 to 2.5 cm long.

Pollen : Pollen loads are light green. The pollen grains are 40 um in diameter. The pollen grains are oval-irregular in shape with 6 apertures which are furrows. The surface structure is netted or pitted and the exine is of medium thickness. The surface of the apertures has scattered granules or projections.

Nectar presentation : Not a significant crop - Aromatic flavour.

Preferred habitat : Dry grasslands on chalky, sandy soils. It thrives in full sun and is resistant to drought.

Particular special features : Honey bees visit for the nectar and pollen. Bumblebees and solitary bees are also attracted.

SEM pictures by Ulrich, S. In: PalDat - A palynological database.
https://www.paldat.org/pub/Hyssopus_officinalis/305570; accessed 2022-03-31 Polar and Equatorial

Marjoram

Family : Lamiaceae
Common Name : Marjoram

Latin Name : *Origanum majorana*
Alternate Name : Pot marjoram, Oregano

Description : A cold-sensitive perennial herb

Flowering times : June - July

Flowers : tiny, two-lipped, tubular, white or pale pink flowers with gray-green bracts bloom in spike-like clusters.

Leaves : are smooth, simple, petiolated, ovate to oblong-ovate, 0.5 - 1.5 cm long, 0.2 - 0.8 cm wide, with obtuse apex, entire margin, symmetrical but tapering base, and reticulate venation.

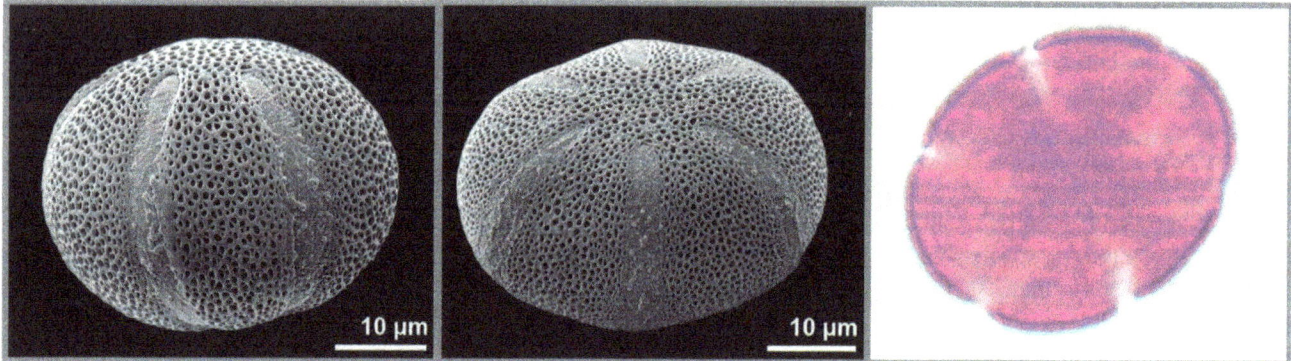

Pollen : Pollen loads are yellow - orange. The pollen grains are 35 um in diameter. The pollen grains are oval-irregular in shape with 6 apertures which are furrows. The surface structure is netted or pitted and the exine is of medium thickness. The surface of the apertures has scattered granules or projections.

Nectar presentation : No honey crop.

Preferred habitat : Rough permanent grassland, hedge banks, scrub, roadsides on dry calcareous soils.

Particular special features : An abundant nectar producer it is a favourite bee plant. Visited by honey bees, solitary bees and bumblebees.

Francis Ratnieks, Professor Of Apiculture (Evolution, Behaviour and Environment) School of Life Sciences at the University of Sussex, rates this herb as one of the most attractive plants for bees in the UK. A must for every bee-lover's garden.

SEM pictures by Halbritter, H. In: PalDat - A palynological database.
https://www.paldat.org/pub/Origanum_vulgare/304448; accessed 2022-03-31 Polar and Equatorial

Catmint

Family :	Lamiaceae	Latin Name :	***Nepeta* spp**
Common Name :	Catmint	Alternate Names :	Catnip
Description :	250 mostly herbaceous perennials	Flowering times :	May - September

Flowers : the tubular flowers can be lavender, blue, white, pink, or lilac, and spotted with tiny lavender-purple dots.

Leaves : opposite heart-shaped, green to grey-green leaves. Usually aromatic foliage.

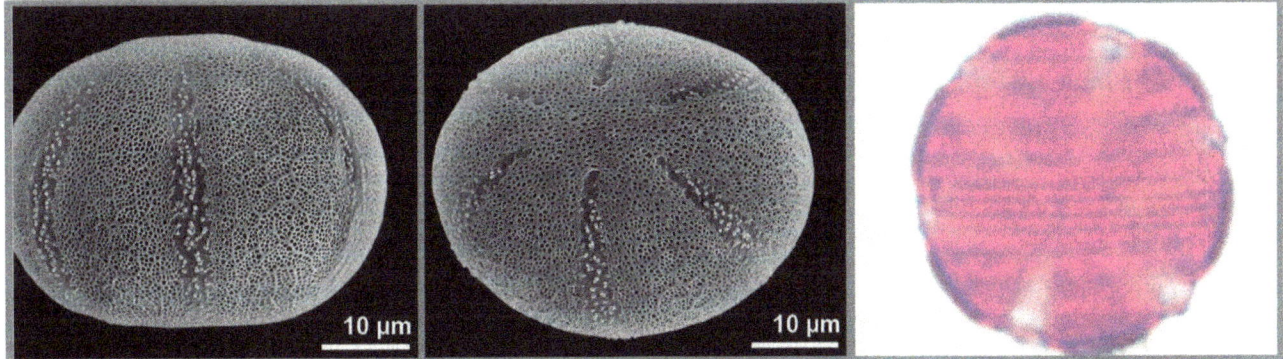

Pollen : Pollen loads are white. The pollen grains are 40 um in diameter. The pollen grains are oval-irregular in shape with 6 apertures which are furrows. The surface structure is netted or pitted and the exine is of medium thickness. The surface in the apertures has scattered granules or projections.

Nectar presentation : No honey crop.

Preferred habitat : Hedge banks, roadsides, rough ground on dry calcareous soils.

Particular special features : Having a short corolla (flower tube) and secreting nectar freely, it attracts honey bees, solitary bees and bumblebees in large numbers.

SEM pictures by Halbritter, H. In: PalDat - A palynological database.
https://www.paldat.org/pub/Nepeta_racemosa/302076; accessed 2022-03-31 Polar and Equatorial

Sage

Family :	Lamiaceae	Latin Name :	*Salvia officinalis*
Common Name :	Sage	Alternate Name :	Garden sage, Common or Culinary sage
Description :	Perennial, woody evergreen shrub	Flowering times :	May - July

Flowers : with lavender flowers most common, though they can also be white, pink or purple.

Leaves : leaves are oblong, ranging in size up to 6.4 cm long by 2.5 cm wide. Leaves are grey-green, rugose on the upper side, and nearly white underneath

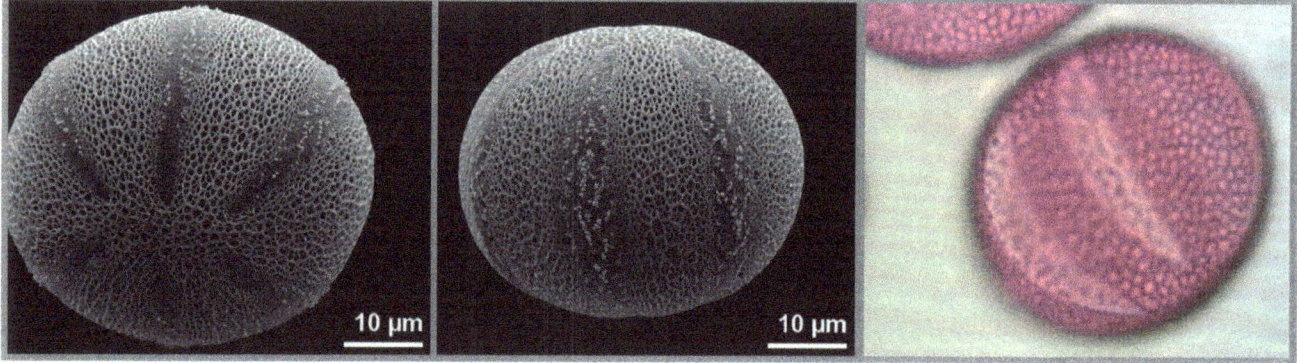

Pollen : Pollen loads are yellow. The pollen grains are 45 um in diameter. The pollen grains are oval-irregular in shape with 6 apertures which are furrows. The surface structure is netted or pitted and the exine is of medium thickness. The surface of the apertures has scattered granules or projections.

Nectar presentation : Not a significant honey crop - Pale yellow to amber in colour. Aromatic flavour and slow granulating.

Preferred habitat : Woodlands, rough grassland, hedgerows, scrub, heaths, rocky hillsides, limestone pavements, sand-dunes and shingle.

Particular special features : Attractive to bumblebees, solitary bees and honey bees.

SEM pictures by Halbritter, H. In: PalDat - A palynological database.
https://www.paldat.org/pub/Salvia_officinalis/304437; accessed 2022-03-31 Polar and Equatorial

Lemon Balm

Family :	Lamiaceae	Latin Name :	*Melissa officinalis*
Common Name :	Lemon balm	Alternate Names :	Balm, Common balm, or Balm mint
Description :	A perennial herbaceous plant	Flowering times :	June - August

Flowers : small white flowers full of nectar, white or light-pink, arranged in axillary whorls.

Leaves : opposite pairs of toothed, ovate leaves growing on square, branching stems. The leaves have a mild lemon scent similar to mint.

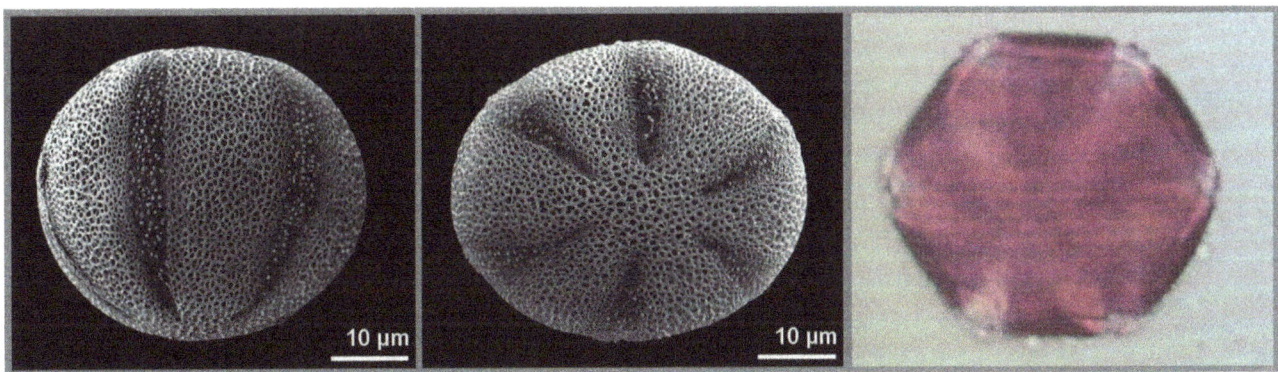

Pollen : Pollen loads are white - grey. The pollen grains are 40 um in diameter. The pollen grains are oval-irregular in shape with 6 apertures which are furrows. The surface structure is netted or pitted and the exine is of medium thickness. The surface of the apertures has scattered granules or projections.

Nectar presentation : No honey crop.

Preferred habitat : Road verges, banks, tips close to habitation.

Particular special features : The flower tube (corolla) length is too great to allow access for honey bees and so it is more attractive to long-tongued bumblebees.

However, it is said that crushing the leaves and rubbing them inside a bait hive makes it more attractive to swarms of honey bees looking for a new home. However, I have no first-hand knowledge of this and it could just be folklore.

The oil from the leaves is called 'Melissa Oil' and it is used in homeopathy and aromatherapy where it is suppoed to have a calming effect.

Melissa officinalis L. - A review of its traditional uses, phytochemistry and pharmacology - Journal of Ethnopharmacology, volume 188, 21 July 2016, Pages 204-228.

SEM pictures by Ulrich, S. In: PalDat - A palynological database.
https://www.paldat.org/pub/Melissa_officinalis/304445; accessed 2022-03-31 Polar and Equatorial

Aquifoliaceae

Key words : Composite flower heads, pitted discs, multiple layers of bracts

The family Aquifoliaceae are a group of about 400 species most of which belong to the genus Ilex, commonly known as the hollies.

The genus is made up of mostly evergreen trees, shrubs, and climbers from the tropics to temperate zones worldwide. However, there are some deciduous trees that lose their leaves at the end of the season.

Most people would identify the holly by its spikey leaves but it is uncommon for hollies to have spikey leaves.

Species of importance to bees :

1. Common Name : Holly
   Latin Name :   *Ilex aquifolium*

Male holly flowers displaying heavily pollen-loaded anthers

Holly

Family :	Aquifoliaceae	Latin Name :	***Ilex aquifolium***
Common Name :	Holly	Alternate Names :	Common holly or English holly
Description :	An evergreen tree or shrub	Flowering times :	May - June

Flowers : are white, four-lobed. Holly is dioecious. In male specimens, the flowers are yellowish and appear in axillary groups. In the female, flowers are isolated or in groups of three and are small and white or slightly pink.

Leaves : leaves are 5-12 cm long and 2-6 cm broad; they are evergreen, dark green on the upper surface and lighter on the underside, oval, leathery and shiny.

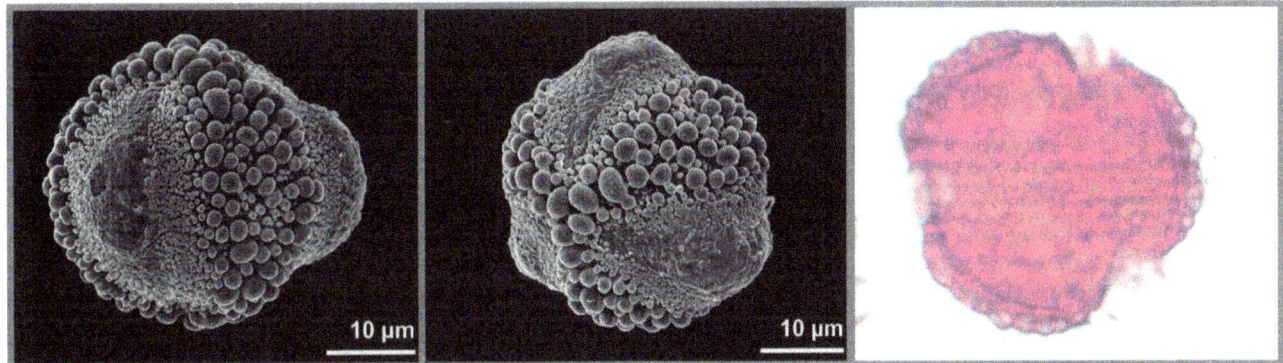

Pollen : Pollen loads are yellow - green. The pollen grains are 35 um in diameter. The pollen grains are oval to round with 3 apertures which are furrows with pores. The surface structure is netted or pitted and the exine is thick with coarse external rods. The are no surface features on the apertures.

Nectar presentation : No honey crop.

Preferred habitat : Woodland on acid soils, wood pasture, scrub and hedgerows.

Particular special features : Good nectar producers they are visited by honey bees, solitary bees and bumblebees.

It has a Pollen Coefficient of 50 which means it is 'normally represented' in honey.

SEM pictures by Halbritter, H. In: PalDat - A palynological database.
https://www.paldat.org/pub/Ilex_aquifolium/306501; accessed 2022-03-31 Polar and Equatorial

Asteraceae - Aster or Sunflower Family

Key words : Composite flower heads, pitted discs, multiple layers of bracts

Asteraceae is a very large and widespread family of flowering plants, The family currently has 32,913 accepted species in 1,911 genera and 13 subfamilies.

In terms of numbers of species, the Asteraceae are rivalled only by the Orchidaceae.

Many members have composite flowers in the form of flower heads surrounded by bracts. When viewed from a distance, this may have the appearance of being a single flower. The sepals are actually bracts, modified leaves, of which there are multiple layers. Each big petal around the flowerhead is actually a complete individual flower called a 'ray flower'.

Pyrrolizidine alkaloids (PAs), sometimes referred to as 'necine bases', are a group of naturally occurring alkaloids produced by plants as a defence mechanism against insect herbivores. PAs have been identified in over 6,000 plants, and about half of them exhibit hepatotoxicity (they are damaging to liver tissue).

They are found frequently in plants in the Boraginaceae, Asteraceae and Fabaceae families and in at least one species in the Lamiaceae.

A rising concern is the health risk associated with the use of medicinal herbs that contain PAs, notably borage leaf, comfrey and coltsfoot in the West, and some Chinese medicinal herbs. This is being taken very seriously in the food industry especially by companies that produce herbal tea infusions.The most commonly known example is in ragwort which is known to harm the liver of horses if left unchecked and gets into the horse's winter feed.

Species of importance to bees :

1. Common Name : Dandelion
   Latin Name : *Taraxacum officinale*

2. Common Name : Ragwort
   Latin Name : *Senecio jacobaea*

3. Common Name : Sunflower
   Latin Name : *Helianthus* spp

4. Common Name : Knapweed
   Latin Name : *Centaurea nigra*

5. Common Name : Sea aster
   Latin Name : *Tripolium pannonicum*

6. Common Name : Dhalia
   Latin Name : *Dhalia* spp

7. Common Name : Tansy
   Latin Name : *Tanacetum vulgare*

8. Common Name : Golden rod
   Latin Name : *Solidago* spp

9. Common Name : Rudbeckia
   Latin Name : *Rudbeckia fulgida*

10. Common Name : Yarrow
    Latin Name : *Achillea millefolium*

Andrena nitida female foraging on a Dandelion

Hoverfly foraging on Ragwort

Honey bee foraging on Tansy

Dandelion

Family : Asteraceae
Common Name : Dandelion

Latin Name : *Taraxacum officinale*
Alternate Name : Common dandelion

Description : A herbaceous perennial plant

Flowering times : March - October

Flowers : the yellow flower heads lack receptacle bracts and all the flowers, which are called florets, are ligulate (shaped like a strap) and bisexual.

Leaves : are 5 - 45 cm long and 1 - 10 cm wide, and are oblanceolate, oblong, or obovate in shape.

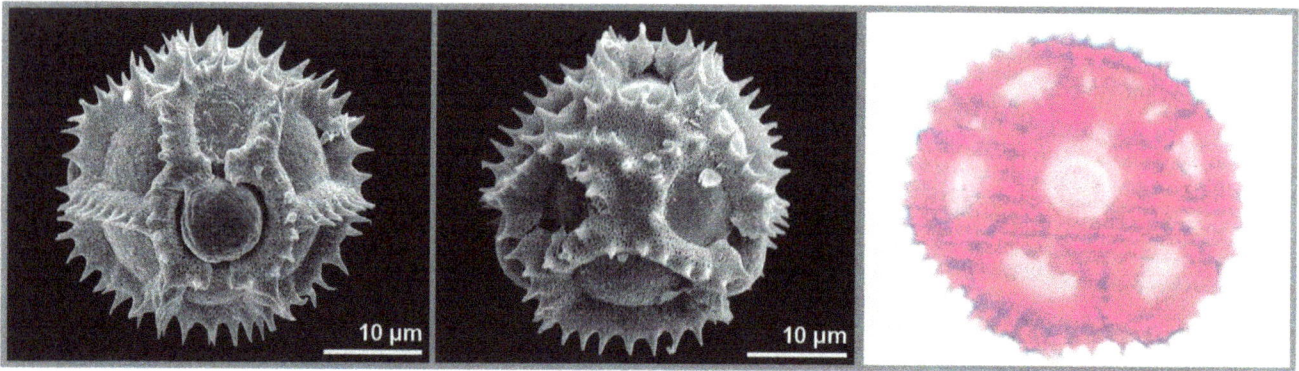

Pollen : Pollen loads are orange. The pollen grains are 30 um in diameter. The pollen grains are irregularly round with 3 apertures which are pores. The surface structure is a combination of a net with spines and the exine consists of small spines. There appears to be no surface features on the apertures.

Nectar presentation : A minor honey crop - Intense golden yellow colour. Rapid coarse hard granulation. Flavour is strong and sharp. Aroma initially repellent like the flower.

Preferred habitat : Chalk grassland, fens, stream sides, sand-dunes and cliffs

Particular special features : A very common sight but a very useful source of nectar and pollen throughout the year. Early in the season when the dandelions are newly out it is a regular occurence to smell the dandelion nectar in honey bee hives and also see the orange staining of the oils from the flower on the combs and in the wax.

The pollen too is vey oily and when making pollen slides needs extra care in de-greasing otheriwse there are obvious yellow oil droplets visible on the slide.

SEM pictures by Bombosi, P. In: PalDat - A palynological database.
https://www.paldat.org/pub/Taraxacum_officinale/305331; accessed 2022-03-31 Polar and Equatorial

Ragwort

Family : Asteraceae
Common Name : Ragwort

Latin Name : ***Senecio jacobaea***
Alternate Names : Common ragwort

Description : A weed of paddocks and gardens
Flowering times : July - October

Flowers : a tall erect plant to 90cm bearing large flat-topped clusters of yellow daisy-like flowers

Leaves : it has finely divided leaves with a basal rosette of deeply-cut, toothed leaves.

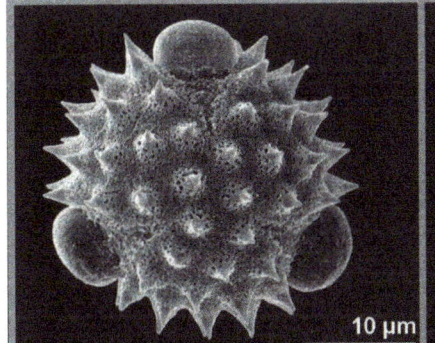 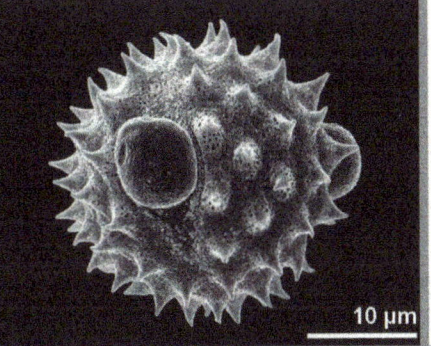 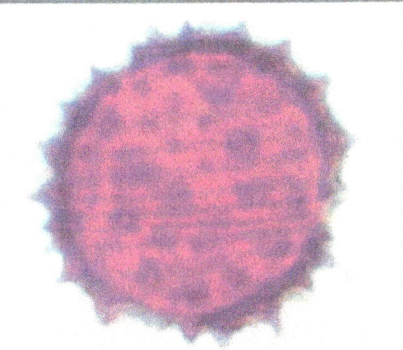

Pollen : Pollen loads are yellow. The pollen grains are 25 um in diameter. The pollen grains are irregularly round with 3 apertures which are pores. The surface structure is a combination of a net with spines and the exine consists of small spines. There appears to be no surface features on the apertures.

Nectar presentation : No honey crop. However, the honey is unpalateable with a strong smell and a bitter taste.

Preferred habitat : Rough grassland, scrub, waste ground, roadsides and sand-dunes.

Particular special features : It is toxic to grazing animals causing liver damage to long-lived animals. However, it is a very good producer of nectar and pollen and attracts a variety of pollinators. This includes honey bees, bumblebees and solitary bees among others.

SEM pictures by Halbritter, H. In: PalDat - A palynological database.
https://www.paldat.org/pub/Senecio_jacobaea/301543; accessed 2022-03-31 Polar and Equatorial

Sunflower

Family : Asteraceae
Common Name : Sunflower
Latin Name : *Helianthus* **spp**
Alternate Name : None

Description : A genus of plants of 70 species
Flowering times : July - September

Flowers : They bear one or more wide, terminal capitula (flower head), with bright yellow ray florets at the outside and yellow or maroon (brown/red) disc florets inside.

Leaves : the petiolate leaves are dentate and often sticky. The lower leaves are opposite, ovate, or often heart-shaped.

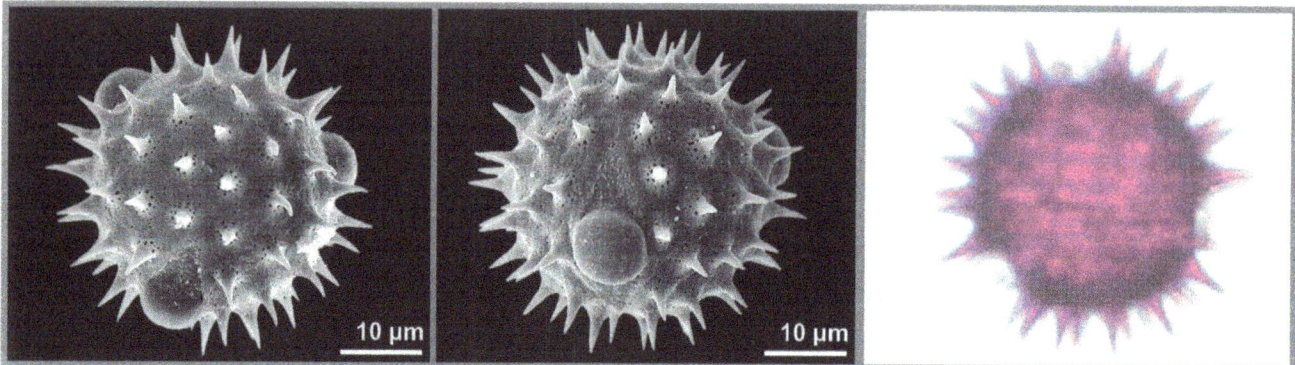

Pollen : Pollen loads are orange. The pollen grains are 35 um in diameter. The pollen grains are irregularly round with 3 apertures which are furrows with pores. The surface structure is isolated dots due to spines and the exine consists of long thin spines. There are no surface features on the apertures.

Nectar presentation : Not a significant honey crop in the UK - Dark yellow colour. Rapid fine soft granulation. Flavour mild but distinctive rather like butter. Fairly strong aroma.

Preferred habitat : Tips and waste places. Widely grown in grdens and as a field crop.

Particular special features : Visited by a wide range of pollinators including honey bees, solitary bees and bumblebees.

It has a Pollen Coefficient of 10 which means it is 'under-represented' in honey.

SEM pictures by Halbritter, H. In: PalDat - A palynological database.
https://www.paldat.org/pub/Helianthus_annuus/304619; accessed 2022-03-31 Polar and Equatorial

Knapweed

Family :	Asteraceae	Latin Name :	*Centaurea nigra*
Common Name :	Knapweed	Alternate Names :	Lesser knapweed, common knapweed
Description :	A perennial herb	Flowering times :	July - September

Flowers : the inflorescence contains a few flower heads, each a hemisphere of black or brown bristly phyllaries. The head bears many small bright purple flowers. Phyllaries are reduced leaf-like structures that form one or more whorls immediately below a flower head.

Leaves : leaves are up to 25 centimetres long, usually deeply lobed, and hairy.

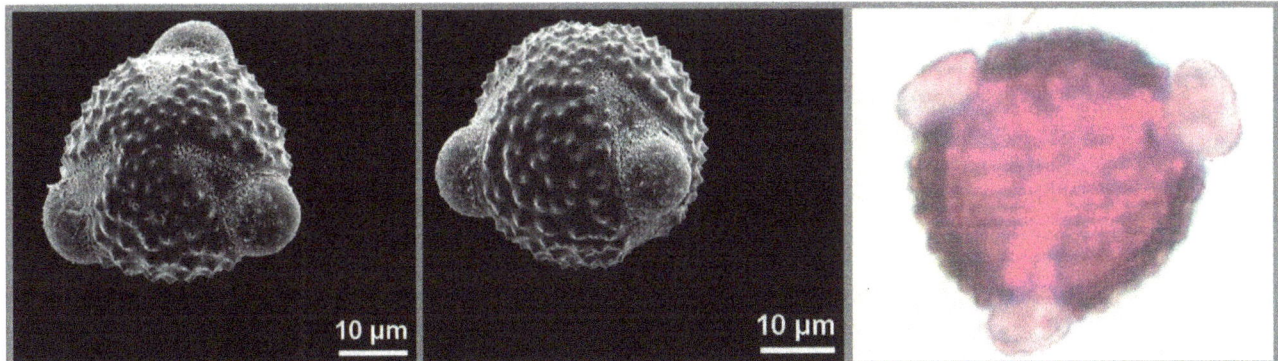

Pollen : Pollen loads are white - grey. The pollen grains are 50 um in diameter. The pollen grains are irregularly round with 3 apertures which are furrows with pores. The surface structure is isolated dots due to spines and the exine is meduim with course external rods. There are no surface features on the apertures.

Nectar presentation : No honey crop.

Preferred habitat : Rough grassland, meadows, pastures, roadsides, sea cliffs and waste ground.

Particular special features : Attractive to honey bees, bumblebees and solitary bees.

SEM pictures by Halbritter, H. In: PalDat - A palynological database.
https://www.paldat.org/pub/Centaurea_jacea/304686; accessed 2022-03-31 Polar and Equatorial

Sea Aster

Family :	Asteraceae	Latin Name :	*Tripolium pannonicum*
Common Name :	Sea aster	Alternate Name :	Seashore aster
Description :	A short-lived perennial herb	Flowering times :	July - October

Flowers : single flower-like 2 - 3 cm capitula surrounded by involucral bracts. Capitula's ray-florets pink or blue (occasionally white)

Leaves : alternate, lowest stalked, upper stalkless.

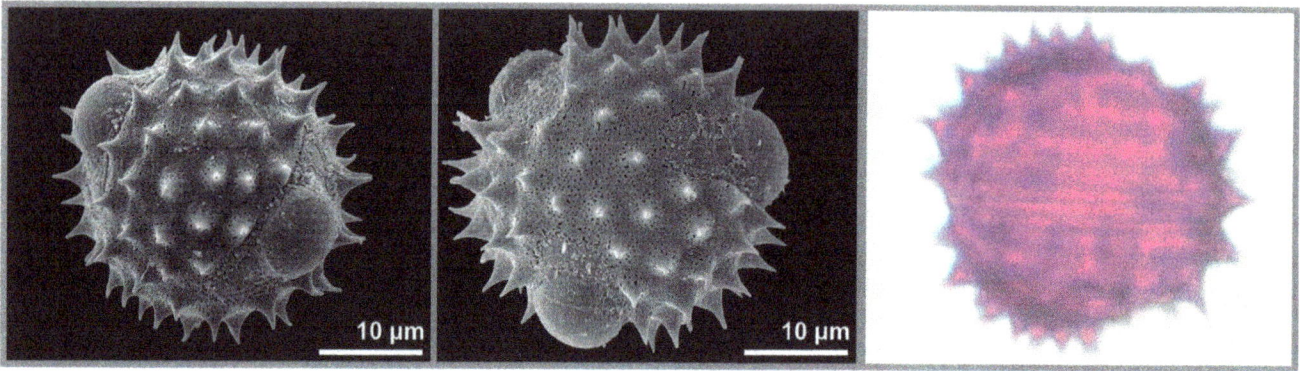

Pollen : Pollen loads are yellow - orange. The pollen grains are 50 um in diameter. The pollen grains are irregularly round with 3 apertures which are furrows with pores. The surface structure is isolated dots due to spines and the exine consists of long thin spines. There are no surface features on the apertures.

Nectar presentation : No honey crop.

Preferred habitat : Salt marshes, creek sides, brackish ditches, tidal rivers and sea cliffs.

Particular special features : A good late source of nectar that sometimes attracts migratory beekeepers for a late crop. In other words, beekeepers may choose to move their bee hives close to the Sea Asters to try to obtain a late crop of honey.

Also attractive to bumblebees and solitary bees.

SEM pictures by Halbritter, H. In: PalDat - A palynological database.
https://www.paldat.org/pub/Tripolium_pannonicum/304606; accessed 2022-03-31 Polar and Equatorial

Dahlia

Family : Asteraceae
Common Name : Dahlia
Description : 42 species of dahlia garden plants

Latin Name : *Dahlia spp*
Alternate Names : None
Flowering times : June - December

Flowers : the flower head is actually a composite with both central disc florets and surrounding ray florets. Each floret is a flower in its own right.

Leaves : the leaves are usually simple, with leaflets that are ovate and 5 - 10 cm long.

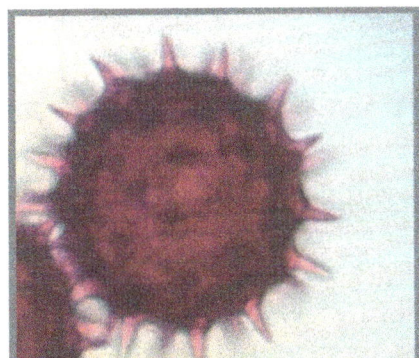 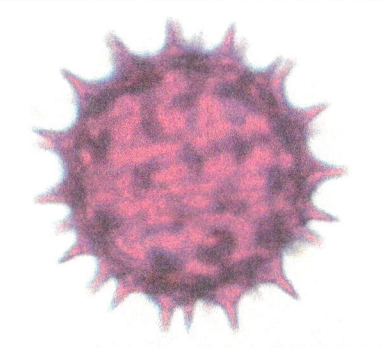 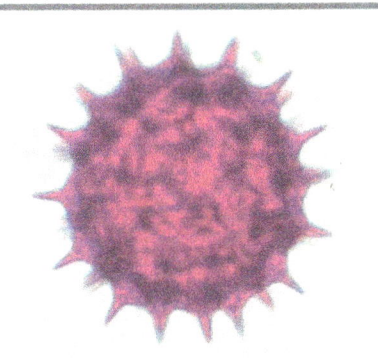

Pollen : Pollen loads are orange. The pollen grains are 40 um in diameter. The pollen grains are irregularly round with 3 apertures which are furrows with pores. The surface structure is isolated dots due to spines and the exine consists of long thin spines. There are no surface features on the apertures.

Nectar presentation : No honey crop.

Preferred habitat : Common garden flowers on well drained soils.

Particular special features : Single varieties are attractive to honey bees, solitary bees and bumblebees.

Tansy

Family :	Asteraceae	Latin Name :	*Tanacetum vulgare*
Common Name :	Tansy	Alternate Name :	Common tansy, Bitter buttons, Cow bitter
Description :	A perennial, herbaceous plant	Flowering times :	July - September

Flowers : the roundish, flat-topped, button-like, yellow flower heads are produced in terminal clusters.

Leaves : are alternate, 10 - 15 cm long and are pinnately lobed, divided almost to the centre into about seven pairs of segments.

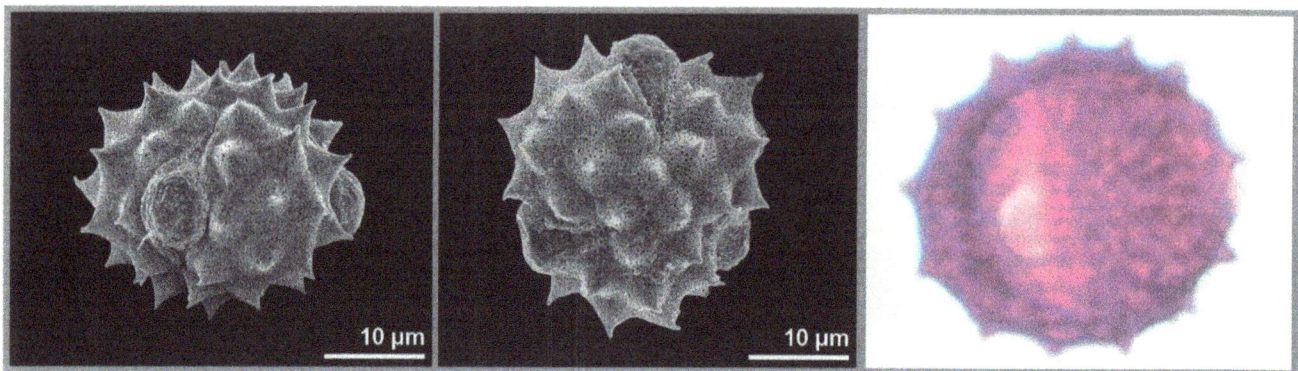

Pollen : Pollen loads are yellow - orange. The pollen grains are 30 um in diameter. The pollen grains are irregularly round with 3 apertures which are furrows with pores. The surface structure is isolated dots due to spines and the exine consists of long thin spines. There are no surface features on the apertures.

Nectar presentation : No honey crop.

Preferred habitat : Rough grassland, roadsides, riverbanks and waste ground.

Particular special features : Very good for solitary bees but also visited by honey bees and bumblebees.

The dried leaves, stalks and flower heads are said to make good smoker fuel.

SEM pictures by Bombosi, P. In: PalDat - A palynological database.
https://www.paldat.org/pub/Tanacetum_vulgare/305188; accessed 2022-03-31 Polar and Equatorial

Golden Rod

Family : Asteraceae
Common Name : Golden rod

Latin Name : *Solidago* spp
Alternate Names : None

Description : A genus of 120 species of plants

Flowering times : July - September

Flowers : The flower heads are usually of the radiate type but sometimes discoid. Floret corollas are usually yellow, but white in the ray florets of a few species

Leaves : the leaf margins are most commonly entire, but often display heavier serration.

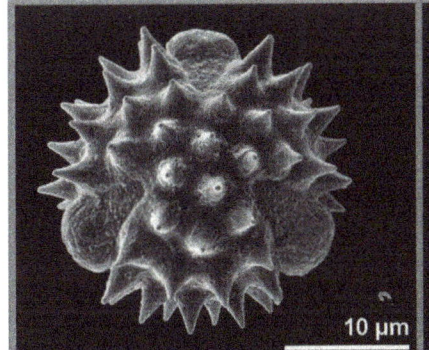 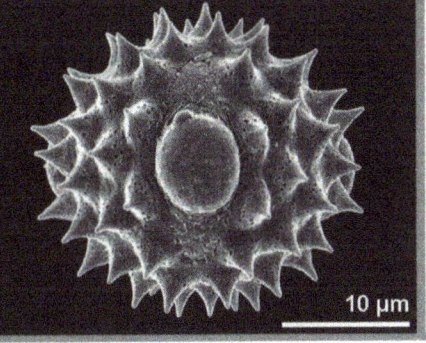 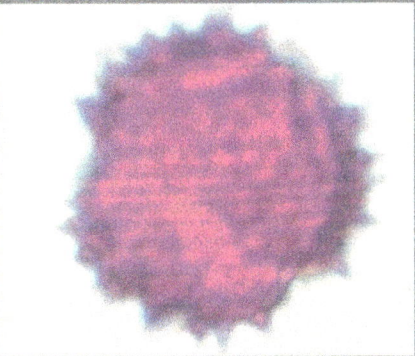

Pollen : Pollen loads are orange. The pollen grains are 25 um in diameter. The pollen grains are irregularly round with 3 apertures which are furrows with pores. The surface structure is isolated dots due to spines and the exine consists of long thin spines. There are no surface features on the apertures.

Nectar presentation : No honey crop.

Preferred habitat : Waste places, roadsides, riverbanks, rough grassland and dunes.

Particular special features : Attractive to a range of pollinators they are readily worked for nectar and pollen. This is a very important source of late nectar and pollen, particulary for preparing honey bee colonies for the winter to come.

SEM pictures by Diethart, B. In: PalDat - A palynological database.
https://www.paldat.org/pub/Solidago_canadensis/304601; accessed 2022-03-31 Polar and Equatorial

Rudbeckia

Family :	Asteraceae	Latin Name :	*Rudbeckia fulgida*
Common Name :	Rudbeckia	Alternate Name :	Coneflowers and Black-eyed-susans
Description :	Herbaceous, mostly perennial plants	Flowering times :	August - October

Flowers : are produced in daisy-like inflorescences, with yellow or orange florets arranged in a prominent, cone-shaped head; "cone-shaped" because the ray florets tend to point out and down as the flower head opens.

Leaves : are spirally arranged, entire to deeply lobed, and 5-25 cm long.

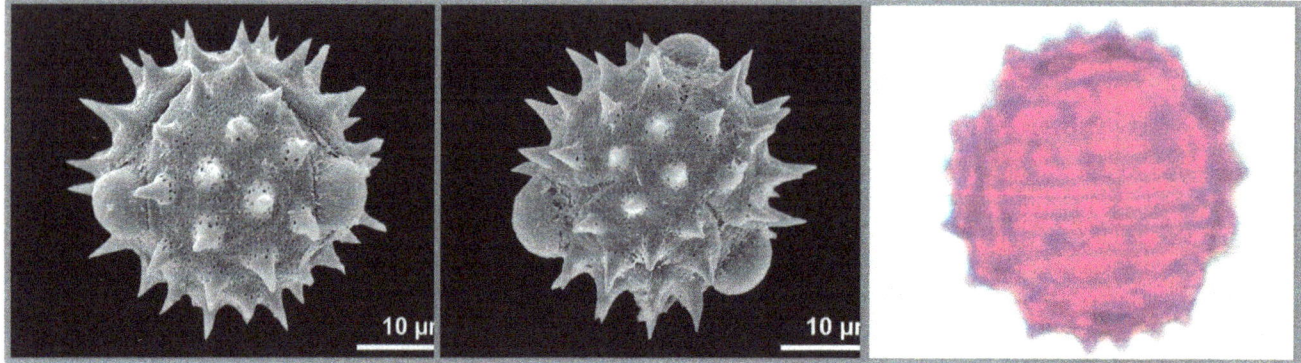

Pollen : Pollen loads are yellow - orange. The pollen grains are 30 um in diameter. The pollen grains are irregularly round with 3 apertures which are furrows with pores. The surface structure is isolated dots due to spines and the exine consists of long thin spines. There are no surface features on the apertures.

Nectar presentation : No honey crop.

Preferred habitat : Dry grasslands, upland forests, particularly in open rocky areas, as well as limestone pavements.

Particular special features : Another good source of late autumn forage for winter stores for honey bees. Also valuable plants for solitary bees and bumblebees.

SEM pictures by Halbritter, H. In: PalDat - A palynological database.
https://www.paldat.org/pub/Achillea_clavenae/305249; accessed 2022-03-31 Polar and Equatorial

Yarrow

Family :	Asteraceae	Latin Name :	*Achillea millefolium*
Common Name :	Yarrow	Alternate Names :	Common yarrow
Description :	An erect, herbaceous perennial	Flowering times :	May - August

Flowers : The inflorescence has 4 to 9 phyllaries and contains ray and disk flowers which are white to pink. The generally 3 to 8 ray flowers are ovate to round.

Leaves : are 5 - 20 cm long, bipinnate or tripinnate, almost feathery, and arranged spirally on the stems

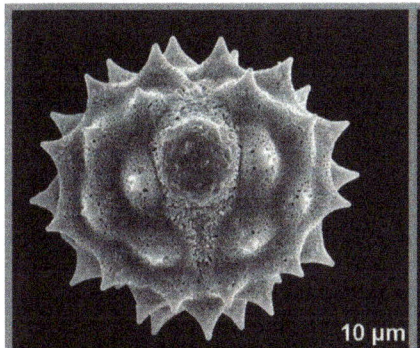 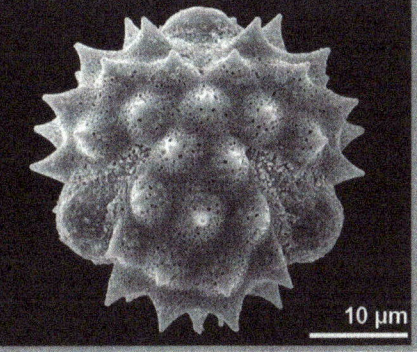 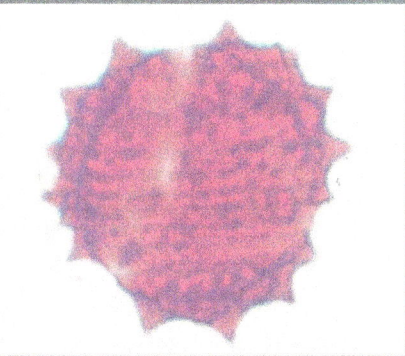

Pollen : Pollen loads are orange. The pollen grains are 30 um in diameter. The pollen grains are irregularly round with 3 apertures which are furrows with pores. The surface structure is isolated dots due to spines and the exine consists of long thin spines. There are no surface features on the apertures.

Nectar presentation : No honey crop.

Preferred habitat : Grassland, lawns, dunes, shingle and wast ground.

Particular special features : Attractive to honey bees, solitary bees and bumblebees.

SEM pictures by Halbritter, H. In: PalDat - A palynological database.
https://www.paldat.org/pub/Achillea_millefolium/304883; accessed 2022-03-31 Polar and Equatorial

Araliaceae - Ginseng Family

Key words : Non-compound umbels with berries, found in moist forests

The Araliaceae is a family made of 52 genera and 700 species of flowering plants including perennial herbs, trees, vines and succulents.

The small greenish-white flowers are regular and may be either bisexual or unisexual.
The flowers have 5 separte sepals and 5 separate petals.
There are usually 5 stamens that alternate with the petals

The ovary is inferior and consists of 2 to 5 carpels. This matures to an even number of chambers and produces a red or purple berry with 1 seed.

Species of importance to bees :

1. Common Name : Ivy
    Latin Name :   *Hedera helix*

Ivy bee foraging on Ivy

Ivy

Family :	Araliaceae	Latin Name :	***Hedera helix***
Common Name :	Ivy	Alternate Names :	Common ivy, English ivy, European ivy
Description :	A rampant, clinging evergreen vine	Flowering times :	September - November

Flowers : individually small, in 3 - 5 cm-diameter umbels, greenish-yellow.

Leaves : are alternate, 50 - 100 mm long, with a 15 - 20 mm petiole.

Pollen : Pollen loads are orange. The pollen grains are 35 um in diameter. The pollen grains are round but triangular in cross-section with 3 apertures which are furrows with pores. The surface structure is netted or pitted and the exine is medium with spaced rods. There are no surface features on the apertures.

Nectar presentation : No significant honey crop - The colour is light and the flavour and the fragrance are repellent. It granulates quickly and very hard.

Preferred habitat : Woods, hedgerows, scrub, walls, cliffs and rock outcrops.

Particular special features : A late source of nectar and pollen and is visited by a wide variety of pollinators. If you are able to watch a patch of ivy on a sunny afternoon it will be alive with flies, wasps, bees and hornets.

The beekeeper aims to get the winter stores around the brood in the bee hive before the ivy is in flower. The beekeeper desires to get the bees to use the ivy nectar as it comes in rather than storing it. This is because it sets very hard in the comb and it is more effort for the bees to use the solid stores when they need it.

There is a species of bee that restricts its pollen intake to ivy alone. This is the ivy bee - *Colletes hederae*

SEM pictures by Halbritter, H. In: PalDat - A palynological database.
https://www.paldat.org/pub/Hedera_helix/304248; accessed 2022-03-31 Polar and Equatorial

Apiaceae - Carrot Family

Key words : Compound umbels with hollow flower stalks and sheathed leaves

Apiaceae is a family of mostly aromatic flowering plants named after the type genus Apium and commonly known as the celery, carrot or parsley family.

It is the 16th-largest family of flowering plants, with more than 3,700 species in 434 genera including well-known and economically important plants.

Most Apiaceae are annual, biennial or perennial herbs (frequently with the leaves aggregated toward the base), though a minority are woody shrubs or small trees

Notice that all the stems of the flower cluster radiate from a single point at the end of a stalk - like an umbrella.

The flowers have 5 sepals, 5 petals and 5 stamens.

The ovary is inferior and consists of 2 united carpels with the same number of styles.
The ovary matures into a dry fruit that splits into individual one-seeded carpels when dry.

Species of importance to bees :

1. Common Name : Wild Carrot
    Latin Name : **Daucus carota**

2. Common Name : Lovage
    Latin Name : **Levisticum officinale**

3. Common Name : Garden Angelica
    Latin Name : **Angelica archangelica**

Honey bee foraging on Lovage

Honey bee foraging on Garden angelica

Wild Carrot

Family :	Apiaceae	Latin Name :	*Daucus carota*
Common Name :	Wild Carrot	Alternate Names :	Bishop's or Queen Anne's lace
Description :	A herbaceous, biennial plant	Flowering times :	June - September

Flowers : The flowers are small and dull white, clustered in flat, dense umbels. The umbels are terminal and 8-10 cm wide. They may be pink in bud and may have a reddish or purple flower in the centre of the umbel. However, this is not reliable and is not used in the flower keys.

Leaves : are tripinnate, finely divided and lacy, and overall triangular in shape.

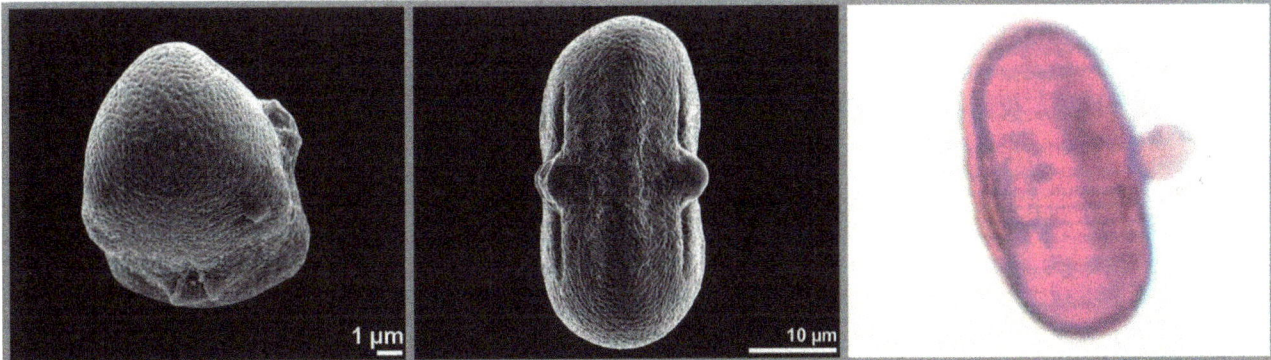

Pollen : Pollen loads are yellow. The pollen grains are 10 x 30 um. The pollen grains are elongated and triangular in cross-section with 3 apertures which are furrows with pores. The surface structure is smooth or indefinite and the exine is thin. There are no surface features on the apertures.

Nectar presentation : Not a significant honey crop in the UK. Dark yellow colour, distinctive flavour and fragrant aroma like the plant.

Preferred habitat : Broken turf, rough grassland, roadsides, waste ground on dry calcareous soils.

Particular special features : The nature of the flowers, being flat and not very deep, means that the nectar is available for a wide range of visitors. This is particulary true for solitary bees.

SEM pictures by Halbritter, H. In: PalDat - A palynological database.
https://www.paldat.org/pub/Daucus_carota/305987; accessed 2022-03-31 Polar and Equatorial

Lovage

Family :	Apiaceae	Latin Name :	*Levisticum officinale*
Common Name :	Lovage	Alternate Name :	None
Description :	Erect, herbaceous, perennial plant	Flowering times :	May - June

Flowers : being produced in umbels at the top of the stems. The flowers are yellow to greenish-yellow, 2 - 3 mm diameter, produced in globose (spherical) umbels up to 10 - 15 cm diameter

Leaves : a basal rosette of leaves and stems with further leaves. The larger basal leaves are up to 70 cm long.

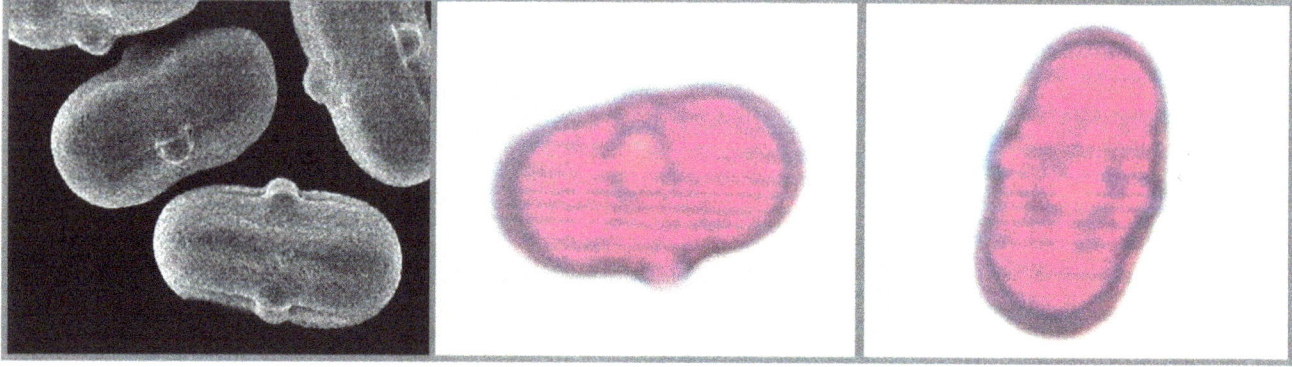

Pollen : Pollen loads are yellow. The pollen grains are 30 um in diameter. The pollen grains are elongated and circular in cross-section with 3 apertures which are furrows with pores. The surface structure is smooth or indefinite and the exine is thin. There are no surface features on the apertures.

Nectar presentation : No honey crop.

Preferred habitat : Rough ground

Particular special features : Attractive to honey bees, solitary bees and bumblebees.

SEM pictures by Halbritter, H. In: PalDat - A palynological database.
https://www.paldat.org/pub/Daucus_carota/305987; accessed 2022-03-31 Multiple grains

Garden Angelica

Family : Apiaceae
Common Name : Garden Angelica

Latin Name : **Angelica archangelica**
Alternate Names : Wild celery, and Norwegian angelica.

Description : A biennial plant to 1.5 m tall

Flowering times : May - July

Flowers : are small and numerous, white, yellowish or greenish, are grouped into large, globular umbels

Leaves : comprise numerous small leaflets divided into three principal groups, each of which is again subdivided into three lesser groups.

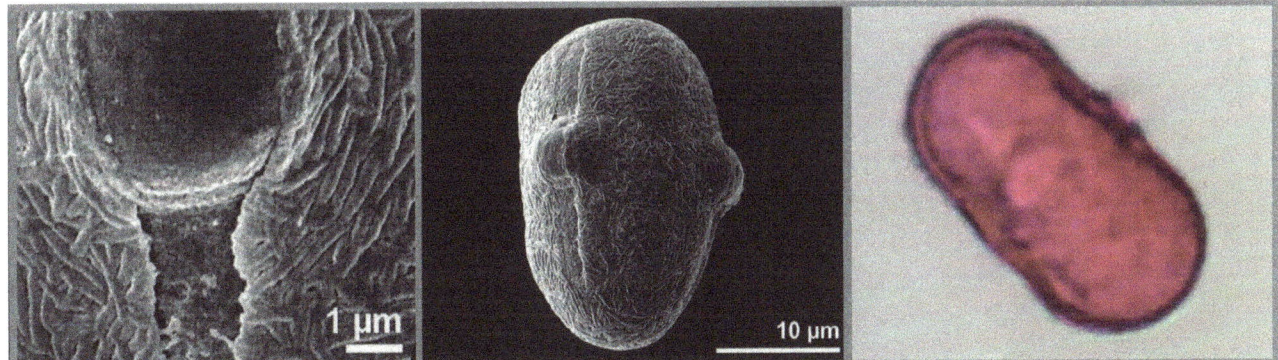

Pollen : Pollen loads are white. The pollen grains are 30 um in diameter. The pollen grains are elongated and circular in cross-section with 3 apertures which are furrows with pores. The surface structure is smooth or indefinite and the exine is thin. There are no surface features on the apertures.

Nectar presentation : Not a significant honey crop in the UK - Dark reddish in colour.

Preferred habitat : Wet meadows, marshes, fens, wet woodlands, ditches, streams, river and lake margins.

Particular special features : Attractive to honey bees, solitary bees and bumblebees.

SEM pictures by Halbritter, H. In: PalDat - A palynological database.
https://www.paldat.org/pub/Angelica_archangelica/301480; accessed 2022-03-31 Equatorial and Detail of aperture

Iridaceae - Iris Family

Key words : Lily-like flowers with leaves in a flat plane

Iridaceae is a family of plants in order Asparagales, taking its name from the irises, meaning rainbow, referring to its many colours.

There are 66 accepted genera with a total of c. 2244 species worldwide. It includes a number of other well known cultivated plants.

The flowers are regular and bisexual with parts in multiples of 3. There are 3 sepals coloured to look like petals, 3 true petals and 3 stamens.

The ovary is inferior and consists of 3 united carpels and 3 stigmas.
It matures as a capsule containing many seeds.

Species of importance to bees :

1. Common Name : Crocus
    Latin Name :   *Crocus* spp

Crocus flower showing the bright orange pollen on the 3 anthers

Crocus

Family :	Iridaceae	Latin Name :	***Crocus* spp**
Common Name :	Crocus	Alternate Names :	None.
Description :	Comprising 90 species of perennials	Flowering times :	February - March

Flowers : cup-shaped, solitary flowers. Their colours vary enormously, although lilac, mauve, yellow, and white are predominant.

Leaves : the grass-like, ensiform leaf shows generally a white central stripe along the leaf axis.

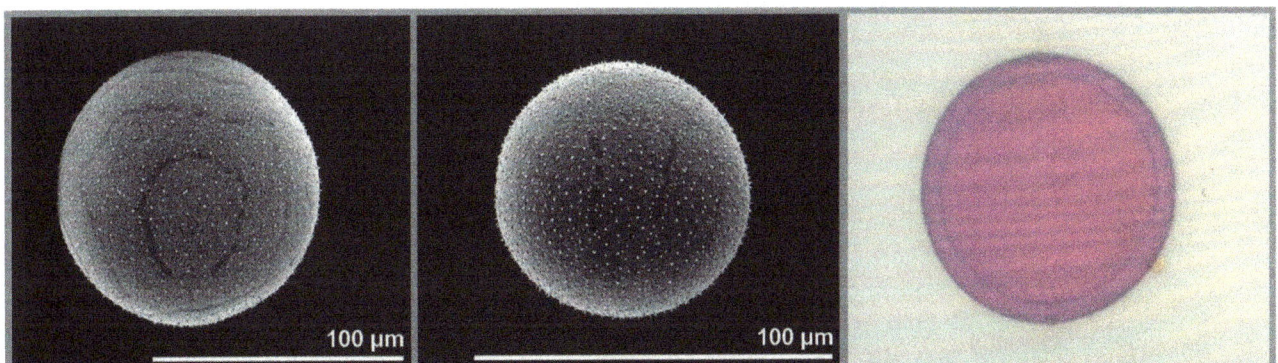

Pollen : Pollen loads are orange. The pollen grains are 75 - 100 um in diameter. The pollen grains are round without any pores. The surface structure is granular with the exine being covered in very small spines or warts. The intine appearing very thick.

Nectar presentation : No honey crop.

Preferred habitat : Roadsides, parks and churchyards.

Particular special features : A very good early source of pollen much visited by honey bees. Also visited, to a lesser extent, by bumblebees and solitary bees.

The pollen loads are bight orange and oily. They need a lot of degreasing to prepare for a pollen slide.

EM pictures by Halbritter, H. In: PalDat - A palynological database.
https://www.paldat.org/pub/Crocus_speciosus/305358; accessed 2022-03-31 Polar and Equatorial

Alliaceae - Onion Family

Key words : Monocots with onion-like bulbs, juicy leaves and small flowers grouped in an umbel

Typically perennial plants that re-grow each year from a bulb. The flowers are grouped in an umbel or sometimes solitary. Individual flowers are lilly-like with 3 sepals and 3 petals that are identical in size and colour.

Most species have 6 stamen with the ovary either inferior or superior maturing as a capsule wih multiple seeds per chamber.

The largest genera are Allium (260-690 species), Nothoscordum (25), and Tulbaghia (22).
Note : The snwdrop used to be a member of this family.

Species of importance to bees :

1. Common Name : Chives
   Latin Name :   **Allium schoenoprasum**

Amaryllidaceae - Amaryllis Family

Key words : Monocots with onion-like bulbs, juicy leaves and flower heads wrapped in a bract

Typically perennial plants that re-grow each year from a bulb. The flowers are grouped in an umbel or sometimes solitary, with a flower head wrapped in a spathe-like bract.

Most species have 6 stamen with the ovary superior maturing as a capsule wih multiple seeds per chamber.
Note : The snwdrop used to be a member of the Alliaceae family.

Species of importance to bees :

1. Common Name : Snowdrop
   Latin Name :   **Galanthus nivalis**

Snowdrops in a woodland setting - Shugborough estate - Staffordshire

Chives

Family :	Alliaceae	Latin Name :	*Allium schoenoprasum*
Common Name :	Chives	Alternate Names :	None
Description :	Bulb-forming herbaceous perennial	Flowering times :	May - August

Flowers : are pale purple, and star-shaped with six petals, 1 - 2 cm wide, and produced in a dense inflorescence of 10 - 20 together.

Leaves : grass-like, leaves, which are shorter than the scapes, are also hollow and tubular.

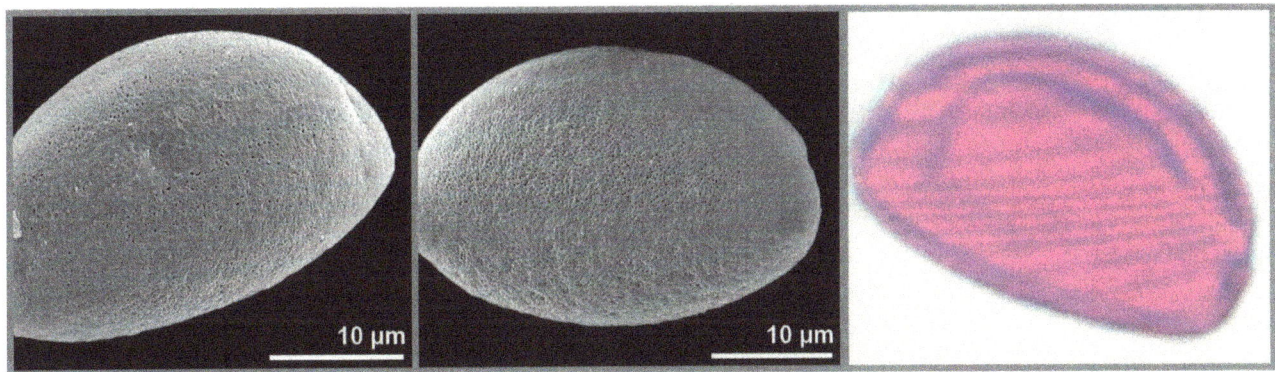

Pollen : Pollen loads are yellow. The pollen grains are 30 x 10 um. The pollen grains are elongated ovals with a single aperture which is a furrow. The surface structure is smooth or indefinite and the exine is thin. There are no surface features on the aperture.

Nectar presentation : No honey crop. If honey is produced then an onion taste and smell is present initially but fades over time.

Preferred habitat : Rocky hill pastures, coastal grasslands on limestone or basic igneous rocks.

Particular special features : Visited for nectar and pollen by honey bees, solitary bees and bumblebees.

SEM pictures by Halbritter, H. In: PalDat - A palynological database.
https://www.paldat.org/pub/Allium_senescens/301397; accessed 2022-03-31 Polar distal and Polar proximal

Snowdrop

Family :	Amaryllidaceae.	Latin Name :	*Galanthus nivalis*
Common Name :	Snowdrop	Alternate Name :	Common snowdrop
Description :	Perennial, herbaceous bulbous plants	Flowering times :	January - March

Flowers : solitary, pendulous, bell-shaped white flower, held on a slender pedicel.

Leaves : produces two linear, or very narrowly lanceolate, greyish-green leaves

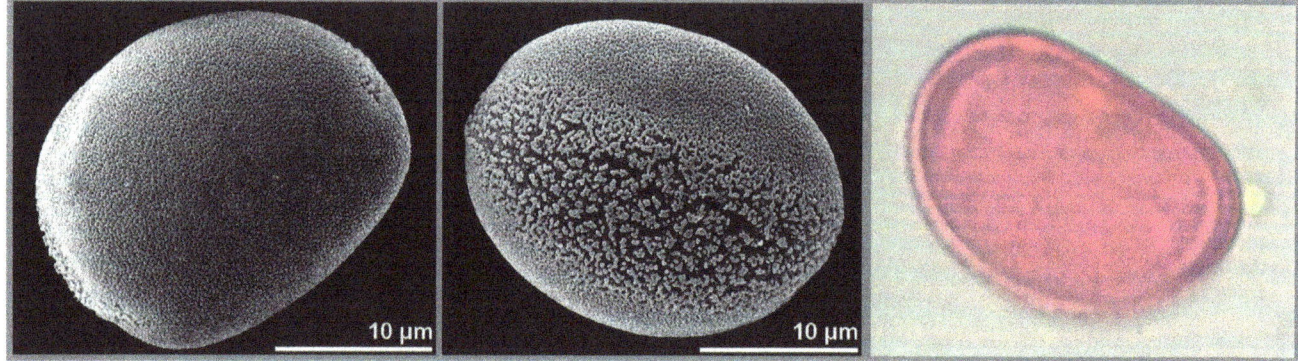

Pollen : Pollen loads are orange. The pollen grains are 30 x 20 um. The pollen grains are elongated ovals with a single aperture which is a furrow. The surface structure is smooth or indefinite and the exine is thin. There are no surface features on the aperture.

Nectar presentation : No honey crop.

Preferred habitat : Woodlands, hedge banks and churchyards.

Particular special features : A very good early honey bee plant producing pollen only.

SEM pictures by Halbritter, H. In: PalDat - A palynological database.
https://www.paldat.org/pub/Galanthus_nivalis/304774; accessed 2022-03-31 Polar and Equatorial

Asparagaceae - Asparagus Family

Key words : rhizomes or bulbs. Very variable - mostly 6 petaled but some exceptions and many have lily-like leaves.

Asparagaceae is a family of flowering plants, placed in the order Asparagales of the monocots. Asparagaceae includes 114 genera with a total of circa 2900 known species.

Distributed nearly worldwide, the family is extremely diverse, and its members are united primarily by genetic and evolutionary relationships rather than morphological similarities.

Species of importance to bees :

1. Common Name : Bluebell
   Latin Name :   *Hyacinthoides non-scripta*

2. Common Name : Asparagus
   Latin Name :   *Asparagus officinalis*

A ladybird eating a caterpillar on Asparagus

Bluebell

Family :	Asparagaceae	Latin Name :	*Hyacinthoides non-scripta*
Common Name :	Bluebell	Alternate Name :	Common bluebell, harebell
Description :	Bulbous perennial woodland plant	Flowering times :	April - May

Flowers : An inflorescence of 5 - 12 (exceptionally 3 - 32) flowers is borne on a stem up to 50 mm tall, which droops towards the tip the flowers are arranged in a 1-sided nodding raceme. Each flower is 14 - 20 mm long.

Leaves : produces 3 - 6 linear leaves, all growing from the base of the plant, and each 7 - 16 mm wide.

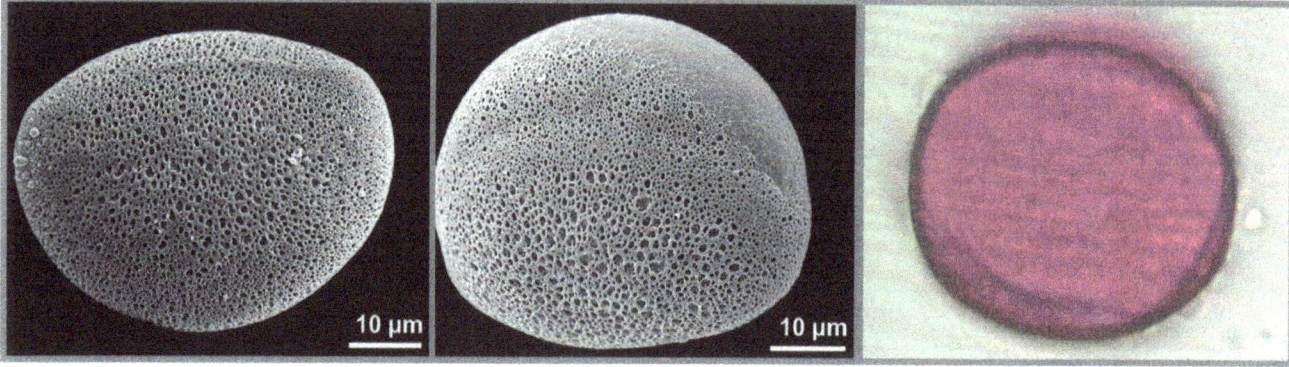

Pollen : Pollen loads are blue. The pollen grains are 30 x 50 um. The pollen grains are elongated ovals with a single aperture which is a furrow. The surface structure is smooth or indefinite and the exine is thin. There are no surface features on the aperture and the intine appears thick.

Nectar presentation : No honey crop.

Preferred habitat : Often forming extensive carpets in deciduous woodland.

Particular special features : A long corolla or flower tube restricts access to the nectar to long-tongued bumblebees, which are usually queens when the bluebells are in flower.

However, honey bees and short-tongued bumblebees are still able to access the nectar by putting their tongues in the seam between the base of the petals in a process known as 'base working'.

Pollen is also foraged by honey bees, short and long-tongued bumblebees and solitary bees.

SEM pictures by Halbritter, H. In: PalDat - A palynological database.
https://www.paldat.org/pub/Hyacinthoides_hispanica/300049; accessed 2022-03-31 Polar and Equatorial

Asparagus

Family :	Asparagaceae	Latin Name :	*Asparagus officinalis*
Common Name :	Asparagus	Alternate Names :	Garden asparagus, Sparrow grass
Description :	A spring vegetable, perennial plant	Flowering times :	June - July

Flowers : are bell-shaped, greenish-white to yellowish, 4.5 - 5.5 mm long. Usually dioecious.

Leaves : are in fact needle-like cladodes (modified stems) in the axils of scale leaves; they are 6 - 32 mm long and 1 mm wide.

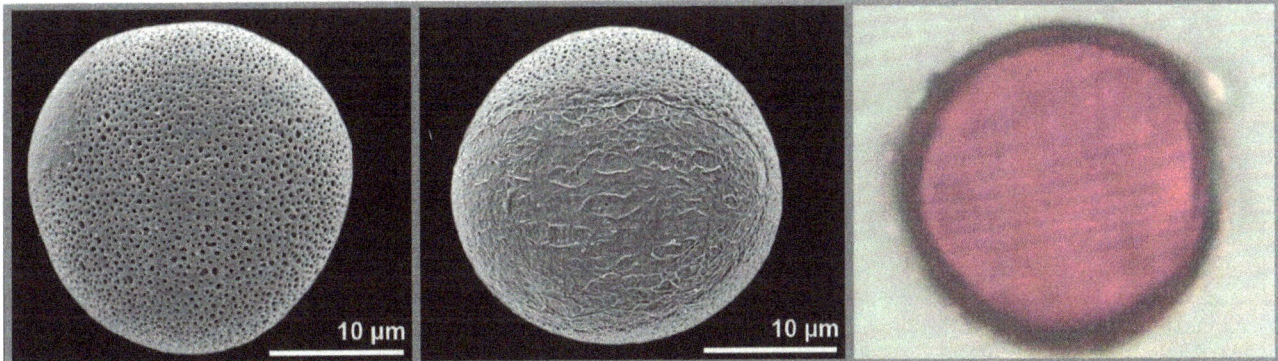

Pollen : Pollen loads are orange. The pollen grains are 25 um in diameter. The pollen grains are elongated ovals with a angle aperture which is a furrow. The surface structure is smooth or indefinite and the exine is of medium thickness. There are no surface features on the aperture and the intine appears thick.

Nectar presentation : No honey crop.

Preferred habitat : Sandy heaths and dunes.

Particular special features : Very attractive to honey bees, bumblebees and solitary bees obtaining both nectar and pollen.

SEM pictures by Halbritter, H. In: PalDat - A palynological database.
https://www.paldat.org/pub/Asparagus_officinalis/304760; accessed 2022-03-31 Polar distal and Polar proximal

Berberidaceae - Barberry Family

Key words : Floral parts in 3s and often several layers of sepals and petals

The Berberidaceae are a family of 18 genera of flowering plants commonly called the barberry family. The family contains about 700 known species, of which the majority are in Berberis. The species include trees, shrubs and perennial herbaceous plants.

The flowers are bisexual and regular and often bloom early in the spring. The flowers are usually 6 true sepals and 6 petals. There are usually 6 or 9 stamens with the ovary positioned superior consisting of a single carpel. The fruit matures to a berry.

Species of importance to bees :

1. Common Name : Mahonia
   Latin Name : *Mahonia* spp

2. Common Name : Barberry
   Latin Name : *Berberis* spp

Honey bee foraging on winter Mahonia

Mahonia

Family : Berberidaceae
Common Name : Mahonia

Latin Name : ***Mahonia* spp**
Alternate Names : Winter mahonia

Description : Genus of about 70 species of shrubs
Flowering times : November - February

Flowers : flowers in racemes which are 5 - 20 cm long.   Leaves : typically have large, pinnate leaves 10 - 50 cm.

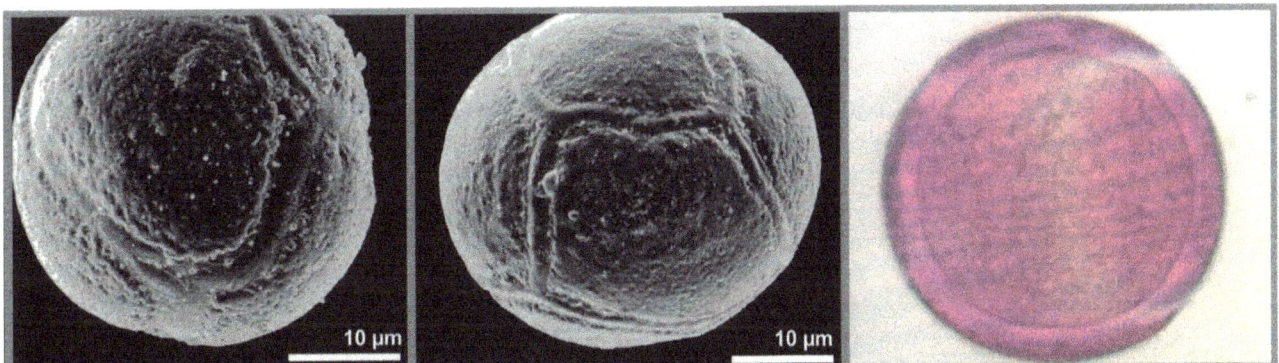

Pollen : Pollen loads are green. The pollen grains are 35 um in diameter. The pollen grains are round with 4 - 6 apertures which are furrows. The surface structure is smooth or indefinite and the exine is of medium thickness. There are no surface features on the apertures.

Nectar presentation : Not a significant honey crop in the UK - Light amber in colour.

Preferred habitat : Hedgerows and woodlands.

Particular special features : A good early source of nectar for honey bees, bumblebees and solitary bees.

This pollen grain is a particular favourite of mine and always reminds me of a tennis ball. The furrows on the pollen grains giving a similar appearance to the tram-lines on a tennis ball.

SEM pictures by Oberschneider, W. In: PalDat - A palynological database.
https://www.paldat.org/pub/Mahonia_aquifolium/303901; accessed 2022-03-31 Polar and Equatorial

Barberry

Family :	Berberidaceae	Latin Name :	*Berberis* spp
Common Name :	Barberry	Alternate Name :	None
Description :	A large genus of shrubs	Flowering times :	April - May

Flowers : are produced singly or in racemes of up to 20 on a single flower-head. They are yellow or orange, 3 - 6 mm long.

Leaves : on long shoots are non-photosynthetic, developed into one to three or more spines.

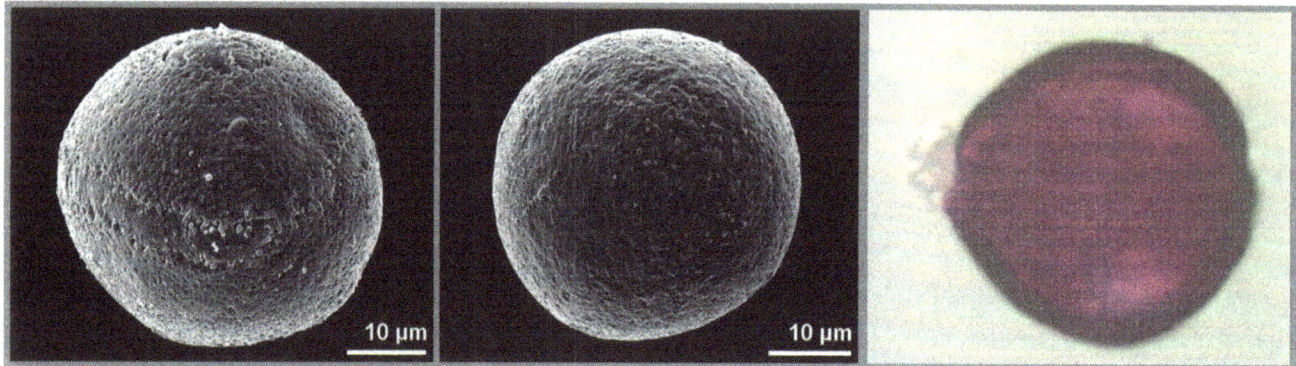

Pollen : Pollen loads are yellow - orange. The pollen grains are 35 um diameter. The pollen grains are round with an indefinite number of apertures which are furrows. The surface structure is smooth or indefinite and the exine is of medium thickness. There are granules or projections scattered on the apertures.

Nectar presentation : No honey crop.

Preferred habitat : Hedgerows, copses and waste ground.

Particular special features : A very good source of early nectar and pollen for honey bees, bumblebees and solitary bees.

SEM pictures by Halbritter, H. In: PalDat - A palynological database.
https://www.paldat.org/pub/Berberis_thunbergii/302268; accessed 2022-03-31 Polar and Equatorial

Lightning Source UK Ltd.
Milton Keynes UK
UKHW051818250223
417660UK00007B/50